The Great Transition to a Green and Circular Economy

Gitte Haar

The Great Transition to a Green and Circular Economy

Climate Nexus and Sustainability

Gitte Haar
Center for Circular Economy
Copenhagen, Denmark

ISBN 978-3-031-49660-8   ISBN 978-3-031-49658-5   (eBook)
https://doi.org/10.1007/978-3-031-49658-5

© The Editor(s) (if applicable) and The Author(s), under exclusive license to Springer Nature Switzerland AG 2024

This work is subject to copyright. All rights are solely and exclusively licensed by the Publisher, whether the whole or part of the material is concerned, specifically the rights of translation, reprinting, reuse of illustrations, recitation, broadcasting, reproduction on microfilms or in any other physical way, and transmission or information storage and retrieval, electronic adaptation, computer software, or by similar or dissimilar methodology now known or hereafter developed.

The use of general descriptive names, registered names, trademarks, service marks, etc. in this publication does not imply, even in the absence of a specific statement, that such names are exempt from the relevant protective laws and regulations and therefore free for general use.

The publisher, the authors, and the editors are safe to assume that the advice and information in this book are believed to be true and accurate at the date of publication. Neither the publisher nor the authors or the editors give a warranty, expressed or implied, with respect to the material contained herein or for any errors or omissions that may have been made. The publisher remains neutral with regard to jurisdictional claims in published maps and institutional affiliations.

This Springer imprint is published by the registered company Springer Nature Switzerland AG
The registered company address is: Gewerbestrasse 11, 6330 Cham, Switzerland

If disposing of this product, please recycle the paper.

Preface

For the past 15 years, I have worked with green transition and circular economy in companies. In these years the world has changed a lot. The need for change in the way we live, the way we consume, and how we run businesses is now clear to most people. The climate is changing with rising temperatures, rapid melting of the ice, burning of forests, and floods. Humans spread and overconsumption harm nature to an extent that will gradually threaten our own habitats and livelihoods. However, I still experience that leaders grope their way forward to embed sustainability strategically. Many business leaders and politicians still believe that sustainability is something that is done alongside the traditional business. This is over now.

I decided to write this book as a guide for companies, organizations, students, and society stakeholders on how to implement the green transition. It explains why it is necessary to regenerate the planet to stabilize the global economies. The markets conditions and legal framework are changing rapidly these years. Demands for sustainability and a transition to a Green and Circular Economy is increasing, by regulators and by consumers.

It is important to understand the links between sustainable living, climate change and the great transition of companies in a holistic way. Companies, private as well as state-owned, are central to the great transition, and the need for a transition is urgent. Action and ambitious strategies are necessary to get started.

I want to contribute to a better planet by bringing stakeholders, management, and citizens to understand why we all must embark on a Great Transition to a Green and Circular Economy. I here introduce the Climate Nexus to describe sustainability holistically. Companies can create profitable business if they understand and embed the complexity of Sustainability. This book is a result of many years of inspirational and innovative work with corporations that led to development of well-tested methods and tools. I hope by sharing these methods and tools and bringing science to the understanding of many to be able to participate in the transition to fair and sustainable planet.

With ambitions on climate neutrality, EU Green Deal, and the Circular Economy Action Plan, the EU is creating completely new market conditions for companies. In the US carbon is becoming a commodity and a market for carbon is emerging rapidly. The Biden administration has put a strong emphasis and a lot of public funds into the transition and technologic development with the Inflation Regulation Act, and the Environmental Justice Law. Still, it seems as if China is approaching the Great Transition at a speed that will put both EU and North America behind.

This book focus on economic and environmental sustainability, but also break with the traditional perception of corporate responsibility in society. The book is also a practical book for companies and other stakeholders on how to ensure competitiveness and create business in a sustainable future. This book has a strong scientific base and will create a deeper understanding of the Green and Circular Economy, climate challenges and the SDGs than what can be achieved from media and general knowledge. The great transition is battleground for nerds.

I hereby thank my family for continuous support in climbing the mountain it was to write this book. A special thanks to Søren Krasilnikoff for being my private editor providing the push and the critics needed. Thanks to all my four children for being the reason and constant inspiration to pursue dramatic change and the courage to speak up.

Copenhagen, Denmark   Gitte Haar

Introduction

A great transition to the Green and Circular Economy is crucial for society and for the planet to protect human existence and to create a fair and sustainable life for all. This transition is the most significant change that societies in the Western World, businesses and consumers will meet in our time. It is a complete transformation of the economy and has huge impact on the future consumption and the future markets. In these years consumers are developing new consumption patterns and demand documented sustainability. The need for change all around society from citizens, consumers, business, and academia is becoming clear. The businesses are the wheels of this transition because they drive consumer behavior, procurement, employee awareness and satisfaction. Companies are increasingly responsible for the full value chains of their business models and on the societies, they built upon.

The overall goal of the great transition to a Green and Circular Economy is to counter the increasing drag on planetary resources and on the climate, creating a sustainable planet for sustainable living of the 9-10 billion people that will inhabit this planet within few decades. A transition of this magnitude will only happen with long-term political visions and the demands from people in strong interaction with the businesses. A global commitment to the transition came with the Sustainable Development Goals (SDGs) presented by UN in 2015. The 17 Sustainable Development Goals is a part of Agenda 2030 for Sustainable Development (sdgs.un.org) and is a shared blueprint for peace and prosperity for people and the planet, now and into the future. It was agreed by state leaders that the SDGs required immediate action by all countries, and almost all UN member countries have committed to the SDGs (UN, www.un.org, 2019). The goals and targets here are to be met by 2030, which is a very short deadline, and we must decide on actions at all levels now.

The EU Commission has transposed the SDGs to the EU Green Deal that was approved in 2020. EU Green Deal are strategies and frameworks of legislation that will drive the change to a Green and Circular Economy supported by the increasing demands from society and consumers.

A Green and Circular Economy is an economy that:

- protects the climate and creates an economy based on renewable energy rather than on fossil energy,
- is based on a Circular Economy and without harvesting virgin raw materials and creating waste,
- transforms the land use (land, oceans, and fresh waters) to regenerative ecosystems for safe and secure food production, a genuine sustainable bioeconomy, and biodiversity to create resilience against the climate changes.
- creates a fair and sustainable living for all.

The increased awareness among people and especially among the young people on human impact on the planet has created a greater willingness to change behavior. Most people do not want a more ascetic, vegan, or minimalistic way of life. Therefore, either raised fingers, minimalism or other kinds of asceticism are not going to save the planet, even though strong movements in this direction are blooming.

The innovative powers and our ability to create technological progress will accelerate the changes necessary to make human activities on the planet sustainable. Along with an increased demand for transparency and sustainability in our value chains. We need a great transition to a genuine sustainable way of living rather than trying to minimize the negative impacts of today's way of living. This the philosophy behind the transition to renewable energy supply, as well as a totally different approach towards nature and how we harvest from nature.

We need to understand that humans are dependent on nature, instead of believing that we can control it. The traditional economic principles as creating sustainable growth, jobs, and export for the companies are still central for the great transition to a Green and Circular Economy to succeed. Now, we must also account for our drag on nature and natural resources, thereby creating the economy that can keep human living within the boundaries of the planet.

This book provides insights, science-based data, detailed visions and roadmaps, and hands-on methods for the Great transition to a Green and Circular Economy, and includes:

Part I of this book gives *the big picture and the background* on the need for change to ensure a livable planet. It gives a *basic understanding of the science behind* the burning platform as well as the historic development that put us on this burning platform of climate change and destruction of wild nature and biodiversity. It also gives an overview of the European legislations from EU Green Deal and gives the background and knowledge to identify the areas where the extensive changes must happen.

Part II introduces the Climate Nexus as a concept where climate impacts and actions are connected to holistic solutions to create genuine, long term sustainable living for humans. Part II introduces the future and how the great transition is formed embedding the main topics:

- Energy transition.
- Transition to a Circular Economy,
- Transition to Sustainable Public and Individual Transport.
- Transition to Sustainable Land Use, Agriculture and Healthy Diets.

Part III includes methods and tools for businesses to transform into sustainable and circular businesses that meet the new EU legislation and drives the changes towards a fair and sustainable planet as so clearly stated in the SDGs. The methods and tools are developed from years of experiences working with corporations in their transition to a Green and Circular Economy, and are all tested with companies.

This book is the first book in English by the same author. Another book is expected on: Rethinking Economics and Business Models for Sustainability in a Nordic perspective and introduces how we behave and manage corporations and society to succeed with this transition. A Case Collection of Nordic corporations is also on its way to illustrate and evaluate how business contribute to the greater transition of society.

About the Author

Gitte Haar is based in Copenhagen (Denmark) and has advised corporations on transforming and preparing their businesses for the green and circular economy for the past 15 years. She is also the author of several books and whitepapers on the green transition and sustainability based on many years of experience with strategic sustainability.

She is a member of non-executive boards of companies working strategically with sustainability, ESG, and the green transition and circular economy. Haar holds an MBA from Copenhagen Business School and a Master of Science (Biology/Agronomy) from the University of Copenhagen, Denmark. She has previously served as an international management consultant at Arthur Andersen and Deloitte.

About This Book

The great transition is the way to a sustainable and fair planet. A transition is necessary to secure supply chains, deliver stable and predictable prices, and ensure access to raw materials in a time of resource scarcity by transforming to sustainable production and consumption.

This book provides broad, essential insights into the main elements of the great transition to Green and Circular economy, and sustainability. Sustainability provide new market conditions for businesses, all over the world. Companies will be subject to new legislation at the corporate and at product level. Companies must meet new requirements and provide a tremendous amount of new ESG data to deliver transparency and traceability in the value chains of products and businesses.

The book aims to close the gaps between science, society, and business. The climate nexus describes the complexity of sustainability and the need for a holistic approach. This book provides solutions, tools and methods to transform today's linear economy and make businesses ready for the future and a society that no longer pushes the planet beyond its limits.

Contents

Part I Introducing the Need for a Green and Circular Economy

1 The Anthropocene Age and a Sustainable Living 3
2 Climate Change and Energy Consumption 11
3 Land Use and the Spread of Humans 25
4 Resource Scarcity 31
5 Green Transition and a New Market Situation 41
6 EU Legislation to a Green Economy 57

Part II Climate Nexus: The Nexus Between Climate Neutrality and Sustainability

7 Introduction to Climate Nexus. 73
8 Energy Transition 75
9 Transition to a Circular Economy. 89
10 Transition to Sustainable Public and Individual Transport 127
11 Transition to Sustainable Land Use, Agriculture and Healthy Diets. 135
12 Driving the Climate Nexus: People and Money 147

Part III Methods and Tools for the Transition to a Circular and Green Economy

13 Introduction to Part III 161
14 Materiality Assessment. 167
15 Changing Company Climate Impacts in Scope 1+2 173

16	**Method to Transform to a Circular Business Model (Scope 3)**	185
17	**Developing a Sustainability Roadmap**	203
18	**Summary**	211

List of Figures

Fig. 1.1	The Antroprocene Age	4
Fig. 1.2	Planetary Boundaries	7
Fig. 2.1	Temperature scenarios from IPCC	13
Fig. 2.2	The Greenhouse Gas Effect	14
Fig. 2.3	Correlation between GHG and temperature	17
Fig. 2.4	UN story map on consequences of Climate Change	19
Fig. 2.5	Large GHG emitters by countries	21
Fig. 2.6	GHG emissions by sector	23
Fig. 3.1	World population	26
Fig. 3.2	Deforestation correlated to population	27
Fig. 3.3	Carbon storage in different natural biotopes	28
Fig. 3.4	Global Land Use	29
Fig. 4.1	Commodity Price Index (CPI)	32
Fig. 4.2	The historic development of the economies	37
Fig. 4.3	Overview of important physical, social and economic dimensions divided into three types of countries	39
Fig. 5.1	Development from CSR to integrated sustainability	43
Fig. 5.2	Green Transition in a company perspective	46
Fig. 5.3	Ellen MacArthur Foundation (EMF) Butterfly Model	49
Fig. 5.4	EU waste generation by economic activity	53
Fig. 6.1	EU Green deal	60
Fig. 6.2	European Sustainability Reporting Standards	62
Fig. 6.3	The framework for implementing Circular Economy in the EU	63
Fig. 7.1	Climate Nexus challenging a fair and sustainable world	74
Fig. 8.1	CO_2 emission by fuel type	76
Fig. 8.2	Energy Prices	80
Fig. 8.3	Global GHG-emission by sector	84
Fig. 8.4	World fossil Carbon Emissions - 1970 to 2018	85

Fig. 8.5	**Worldwide CO_2 emission by region per capita (2017)**	86
Fig. 8.6	**GDP Map per Capita**	87
Fig. 9.1	**Resource hierarchy**	91
Fig. 9.2	**Linear economy**	92
Fig. 9.3	**Saving from recycling or other circular concepts (CO2eq)**	95
Fig. 9.4	**Circular value chain**	97
Fig. 9.5	**Business Models**	102
Fig. 9.6	**EU Target Recycling Rates**	105
Fig. 9.7	**Waste generation in EU by country (2020)**	106
Fig. 9.8	**Material streams to products**	109
Fig. 9.9	**Representation of flows of raw materials and current supply risk**	110
Fig. 9.10	**European Plastic Pact**	116
Fig. 9.11	**New Business Model for Fiber-based paper and cardboard**	118
Fig. 9.12	**Textile value chains**	119
Fig. 9.13	**Circular value chains in construction**	121
Fig. 9.14	**Circular Economy vs. Sharing Economy**	122
Fig. 10.1	**GHG by means of transport**	128
Fig. 11.1	**EU Biodiversity strategy**	137
Fig. 11.2	**EU Farm to Fork**	138
Fig. 11.3	**Population by Region (1873–2100)**	140
Fig. 11.4	**GHG emission in the value chain from Food Products**	141
Fig. 11.5	**The environmental impacts from food production and agriculture**	142
Fig. 12.1	**Climate Nexus**	148
Fig. 12.2	**Stakeholders in a green and circular economy**	149
Fig. 12.3	**Consumer patterns hierarchy**	151
Fig. 13.1	**UN Climate Protocol for companies**	163
Fig. 14.1	**Materiality Assessment**	169
Fig. 15.1	**Process for minimizing GHG emissions in scope 1 and 2**	174
Fig. 15.2	**Sustainable Buildings-creating positive impacts**	182
Fig. 16.1	**Process for transforming to a circular business model mitigating scope 3**	190
Fig. 17.1	**Sustainability Roadmap**	204

List of Tables

Table 3.1	**Development of Global Middle Class**	26
Table 3.2	**Development of Global Middle Class consumption**	26
Table 6.2	**Important legislation and strategies from EU on various product groups**	65
Table 8.1	**GHG emissions by electricity source based on Lifecycle Analyses (LCA) to identify the most climate friendly**	77
Table 9.1	**Elements in Circular Economy and Sharing Economy**	123

Part I
Introducing the Need for a Green and Circular Economy

Part I of this book gives *the big picture and the background* on the need for change to ensure a livable planet. It gives a *basic understanding of the science behind* the burning platform and the historic development that put us on this burning platform of climate change and destruction of wild nature and biodiversity. It also gives an overview of the European legislations from EU Green Deal and gives the background and knowledge to identify the areas where the extensive changes must happen.

Chapter 1
The Anthropocene Age and a Sustainable Living

We are now experiencing the consequences of human living and rapid population growth throughout the last 100 years. The spread of humans and the way of living have had enormous impact on the planet—both on land, oceans, and fresh waters.

Climate change is happening all around the world and is due to the emissions of greenhouse gases (GHG) from human activity. The climate is changing much faster than the UN Intergovernmental Panel on Climate Change (UN IPCC) has predicted over the last decades. The widespread human activities have also resulted in a negative impact on biodiversity and wild nature to an extent that scientists call a mass extinction of species. The Earth is running out of material resources to meet people's basic needs. We have entered the Anthropocene Age, where humans as a single species is all-dominant and is defining the geological state of the planet. Geologists name the ages of the planet from large geological or biological events. Now the presence of one species—*Homo sapiens*—is defining the state of planet and its surface.

Humans have become our own worst enemy because we now endanger our own habitats and the resilience that wild nature provides. Human survival and wellbeing depend on natural ecosystems and on a rich and biodiverse existence of plants and animals. Plants, animals, and microorganisms are important for our food supply and are a treasury of biology and genetic variation necessary for food production, medical, and biotechnical development. Wild nature and biodiversity are creating the resilience that humans need against climate change, draughts, and pandemics, as the Corona outbreak. With high population density pandemics will certainly occur again and more lethal ones than the Coronavirus.

Wild nature and biodiversity restore carbon in and above ground and are part of mitigating climate change and greenhouse gas emissions (GHG). As a dominant and self-confident species, we hold an ethical responsibility not to harm the ecosystems we depend upon.

Apart from the environmental and financial potential, there is an unrecognized human potential in a great green and circular transition and a sustainable way of

© The Author(s), under exclusive license to Springer Nature Switzerland AG 2024
G. Haar, *The Great Transition to a Green and Circular Economy*,
https://doi.org/10.1007/978-3-031-49658-5_1

living. In corporations that work with genuine sustainability employees at all levels will experience greater work satisfaction and a greater purpose in their work life. We are facing a crisis of mental health in the old, industrialized part of the world that calls for a larger understanding and integration into nature. Interestingly, studies indicate that people are more environmentally conscious at home than at work. Maybe due to the management's lack of strategic focus and awareness on the potential of a new Green and Circular Economy. This book hopefully contributes to this understanding.

The impacts on environment and on humans from our overconsumption are becoming increasingly clear. Overconsumption and linear business models affect the planet and humans to a larger extent than earlier. Today it is a challenge to make genuine sustainable choices. It is difficult to live sustainably because most products or services are a result of long, untransparent, and untraceable value chains. These long value chains have created a global economy that is unsustainable and economies that are unstable and volatile based on inflation and resource scarcity.

The three elements illustrated in the Anthropocene Age (Fig. 1.1) are intensively affecting each other in a way that accelerates the negative impacts on the planet, on the climate, and on the way we live. Increasing unsustainable harvest of virgin resources will cause increasing climate change which again will increase the distinction of wild nature and biodiversity due to draught, burn, and flooding.

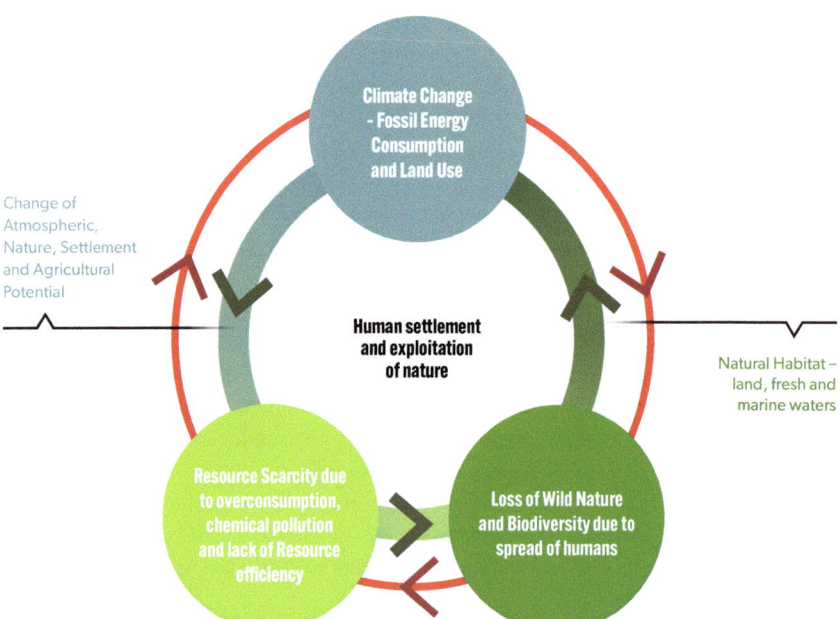

Fig. 1.1 The Antroprocene Age. Illustration of the Anthropocene Age and how the three major elements effect each other negatively today. This must be turned into an accelerating positive impact and genuine sustainability

We need to change this negative spiral into a positive spiral by a great transition into a Green and Circular Economy independent of virgin resources and based on renewable energy, and sustainable and regenerative ecosystems. This requires tremendous changes in human behavior and human consumption where politicians and corporations are essential to drive this change.

Sustainable Development

The dependency on virgin resources and scarcity is not a new challenge. Two European prime ministers realized this already in the 1980s when discussing the large threats against Europe. The Cold War was petering out and the threat of a third nuclear war was less. It was time to look up and investigate the new risks toward Europe. Prime Ministers Margaret Thatcher (UK) and Gro Harlem Brundtland (N) initiated a commission under the UN and Gro Harlem Brundtland who was no longer prime minister in Norway headed the commission that should uncover the potentials of sustainable growth considering human impact on nature, climate, and natural resources. This resulted in the Brundtland Report and a definition of sustainable growth, namely:

> **A sustainable development is a development that meets the needs of the present generation without jeopardizing the needs of future generations for achieving their goals.**
> —UN Brundtland report, 1986.

Our way of living does still not meet a sustainable development, globally—almost 40 years after. On the contrary, we have accelerated unsustainable and linear growth since the report came out. We barely understand how sustainable development impacts our ways of living and the gap to a sustainablity is even bigger today. The industrial, linear way of production based on long, global value chains has become the largest challenge to our economies, to human lives, and to the planet. Especially the consumption and value chains driven by the old, industrialized countries have created untransparent, untraceable, and unstable value chains that feed the unsustainable living, here. The industrialized countries do not hold raw materials to meet their demands nor to continue growth rates. The population growth in these old, industrialized countries is low or negative, but still the consumption and impacts on the planet are increasing here. We need a transition of the Western economies to a Green and Circular Economy rather than adjust linear economic growth.

> Sustainable living and a great transition are necessary to be able to provide for humans and our life on Earth. The planet will survive but humans have created a way of living that is hazardous to our own existence.

Planetary Boundaries

The limits of the planet's carrying capacity have become a widespread topic for discussion and some believe that our way of life today requires more globes. Researchers have investigated the ecological state of the planet and a large group of researchers published the concept on **Planetary Boundaries**, first in 2009 and since in updated versions (see https://www.stockholmresilience.org) (Azote for Stockholm Resilience Centre, 2023 (2009, 2015)). This concept of Planetary Boundaries is to illustrate the impacts of human activity on the planet by the variables in the ecosystems and the atmosphere. In Fig. 1.2 three version of the Planetary Boundaries illustrates that crossing the planetary boundaries is developing rapidly and over the last 14 years from 2009 where 3 boundaries where crossed until now (2023) where six boundaries are crossed. It also shows that we cannot continue to live the way we do if we shall provide for all humans on the planet. The researchers look at nine ecological measures to assess human impacts. Humans are pushing the planet to its boundaries on at least six ecological parameters.

> **Ecology is the study of organisms and how they interact with the surrounding environment, including the relationship between living things and their habitats.**

The boundaries are exceeded on the **biochemical flows** of Phosphorus (P) and Nitrogen (N) because of intensive agriculture, forestry, and aquaculture. Phosphorus and Nitrogen are the most important fertilizing compounds for plants. These nutrients (fertilizers) are washed out in rivers and oceans causing eutrophication of fresh and marine waters resulting in oxygen depletion, death of fish and animals, and have large negative impacts on fresh and marine ecosystems.

Land systems change and **climate change** has also exceeded the planetary boundaries again due to intensive agriculture and emissions from fossil fuels. Climate change is stated with a note on increasing risk of going fully into red. **Novel entities** refer to compounds, as e.g., chemical pollution, that are novel in a geological sense and that could have large-scale impacts that threaten the integrity of Earth system processes. **Biosphere integrity** refers to loss of biodiversity and extinction of species, and is now also in the red, due to the large loss of wild nature. **Ocean acidification** is still within the safe space (green) although this has become a hot topic in the debate on the blue sustainability and loss of coral reefs, overfishing, and loss of ocean mammals. The ocean absorbs lots of CO_2 which causes the acidification. When oceans are no longer a sink of CO_2 climate change will properly accelerate. **Stratospheric ozone depletion** is within the safe space and the emission of gasses that deplete the ozone layer is under increasing control.

We have now pressured the ecosystems by human activities causing climate change, decrease in biodiversity, and destruction of natural ecosystems. All important for human existence and dependency on the planet and atmosphere resilience. The Planetary Boundaries illustrates that impacts from human living have reached

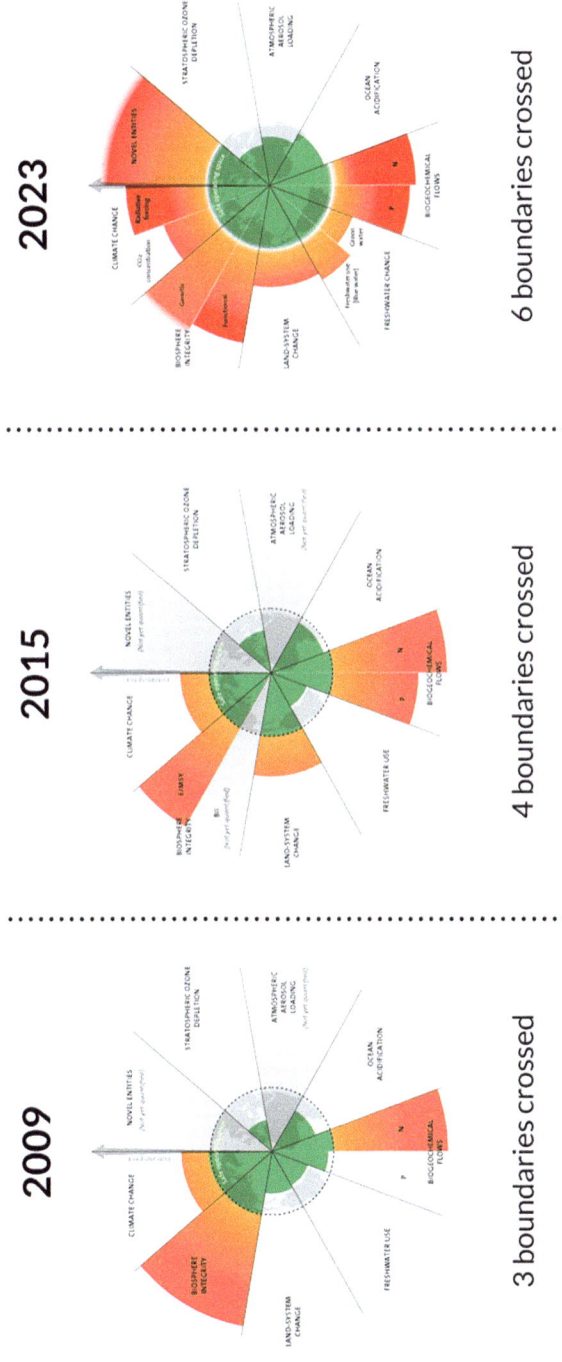

Fig. 1.2 Planetary Boundaries. Illustration of Planetary Boundaries—a method to assess the human impact on the planetary boundaries by Stockholm Resilience Centre

levels where they cannot be recovered—they are irreversible, as the climate changes. Human activities already threaten people's way of life to an extent where we actively must regenerate the ecosystems to ensure a livable planet, and to an extent that will cause migration of people to move away from areas not suited for food production or human living.

The approach and the concept of Planetary Boundaries gave inspiration to the development of the Doughnut Model by Professor Kate Raworth, which has received much attention. Several politicians now want to use this model as a roadmap for sustainable and regenerative economic development in cities around the world. The Doughnut Model is a framework for economic and social development within the planetary boundaries (Raworth, 2018). The Doughnut Model is investigated further in (Haar, Rethink Economics, 2024) and explains how humans need to stay within the Planetary Boundaries (the outer limits of the doughnut) and yet create a decent life with access to healthy food, education, and social welfare and health (the inner limits of the doughnut).

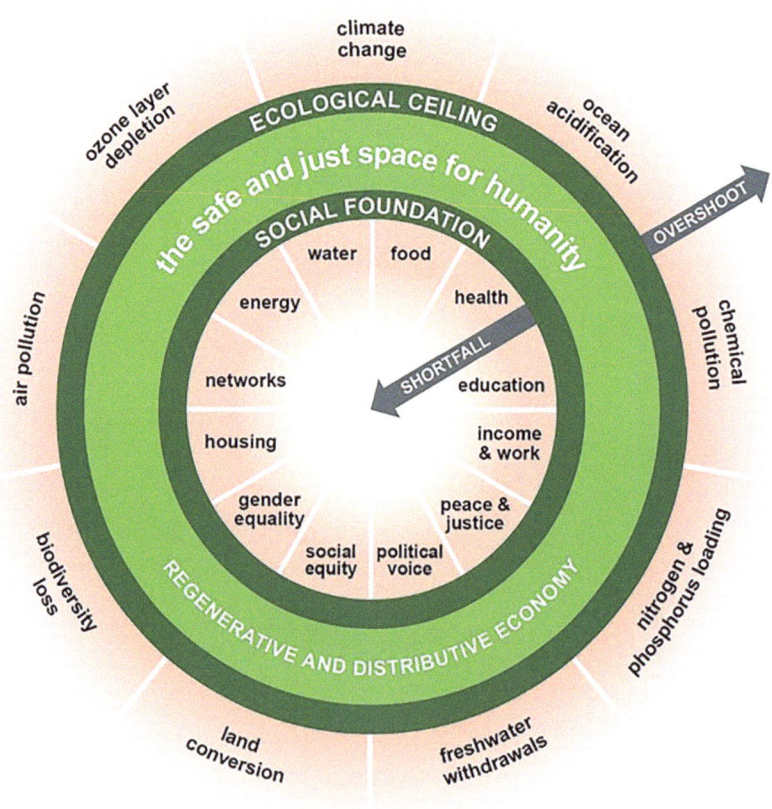

Doughnut Model. Doughnut Model by Kate Raworth, Doughnut Economics

To investigate the backgrounds and potentials for the green and circular transition a description of the science and the historical elements in the development of the Anthropocene Age is included in part I, as (1) climate change and energy consumption, (2) land use and loss of wild nature and biodiversity, and (3) resource scarcity follows here in this introduction. In part II of this book, the solutions are unfolded in the Climate Nexus.

References

Azote for Stockholm Resilience Centre, b. o. (2022 (2015)). Planetary Boundaries. Persson et al and Steffen et al, Stockholm, Sweden

Haar, G. (2024). Rethink Economics and Business Models. *Rethink Economics*. SpringerNature.

Raworth, K. (2018). *Doughnut Economics*. Cornerstone.

Chapter 2
Climate Change and Energy Consumption

Climate Change is a reality, and we all experience flooding, heavy rains, drought, burning of forest and grassland, and desertification of agricultural land, not only in Africa but all around the World. It is a fact that climate change is rapidly changing our cities, agriculture, oceans, and nature. Most people accept that climate change is due to human activity causing increasing temperatures in the atmosphere, the oceans, the ice (cryosphere), and on land.

This chapter reviews Green House Gases (GHGs), the greenhouse effect, and GHG emissions in detail and all data presented here are publicly available and can be found online.

> **Internet pages for more data:**
> - International Energy Agency (IEA.org).
> - Our World in Data (ourworldindata.org).
> - Energy data from the European Commission.
> - The UN has various data sites mainly based on data from IPCC: un.org; unstats.un.org/unsd/envstats/climatechange.cshtml.

Analyses, sites, and articles are often based on data from the UN, EU, or EPA (epa.gov)—the US Environment Agency. Thus, many good and valid data are available online, but make sure to check the original data sources and validity. Pay attention to whether data are comparable if needed, as there are different methods for calculating greenhouse gases and their origin.

Climate Commitments and Paris Agreement

The Paris Agreement is the global UN framework that most countries have committed to. The Paris Agreement (2015) committed the signing countries to limit the rise of temperature from 1.5 to 2 °C. The latest UN report (IPCC, IPCC Sixth Assessment Report, 2022) presents some worrying forecasts on the rise of temperature and the effects of Climate Change. The temperature has already risen to a global average of 1.15 °C (2022) since pre-industrialization (average of 1850–1900) with the last 8 years being the warmest (IPCC, UN Climate Report, 2023). In 2022, large areas with abnormal precipitation were reported from all continents. This year (2023), it seems as if the global average temperature will hit the highest level since pre-industrialization with an increase of 1.8 °C.

On a global level, Green House Gas (GHG) emissions are still rising with only little signs of declining emissions or even flattening out. This is a challenge when looking at the recent forecast scenarios by IPCC (2022). Figure 2.1 shows that to meet the 1.5 and 2 °C commitments, an immediate decline of GHG emissions is needed, also to a larger extent than committed upon. The implemented policies will lead to an expected increase in temperature of 3.2 °C in the range of 2.2–3.5 °C. This will have huge impact on the planet, the ecosystems, and the human habitats.

> **"Global warming, reaching 1.5°C in the near-term, would cause unavoidable increases in multiple climate hazards and present multiple risks to ecosystems and humans (very high confidence)."**—IPCC (2022)

The ambitions to mitigate climate change have become huge over the past few years. Europe, the USA, China, India, and many other countries have set targets on climate neutrality. If climate change and GHG emissions are to stop, it requires major and immediate changes now.

> **The targets of the large emitters here in 2023 are (scopes 1 and 2):**
> - EU climate neutral in 2050—with 55% reduction in 2030. Some EU countries are discussing Climate Neutrality already in 2035.
> - The USA carbon neutral in 2050 and 50% reduction in 2030 (California targets climate zero in 2045, Texas and Florida working against ESG).
> - China climate neutral in 2060.
> - India climate neutral in 2070.

Without a doubt we need dramatic action now. Even so, areas will become unlivable for humans, and unsuitable for food production, as seen in Africa, Southern Europe, California, etc. The negative impact on natural ecosystems is huge and will

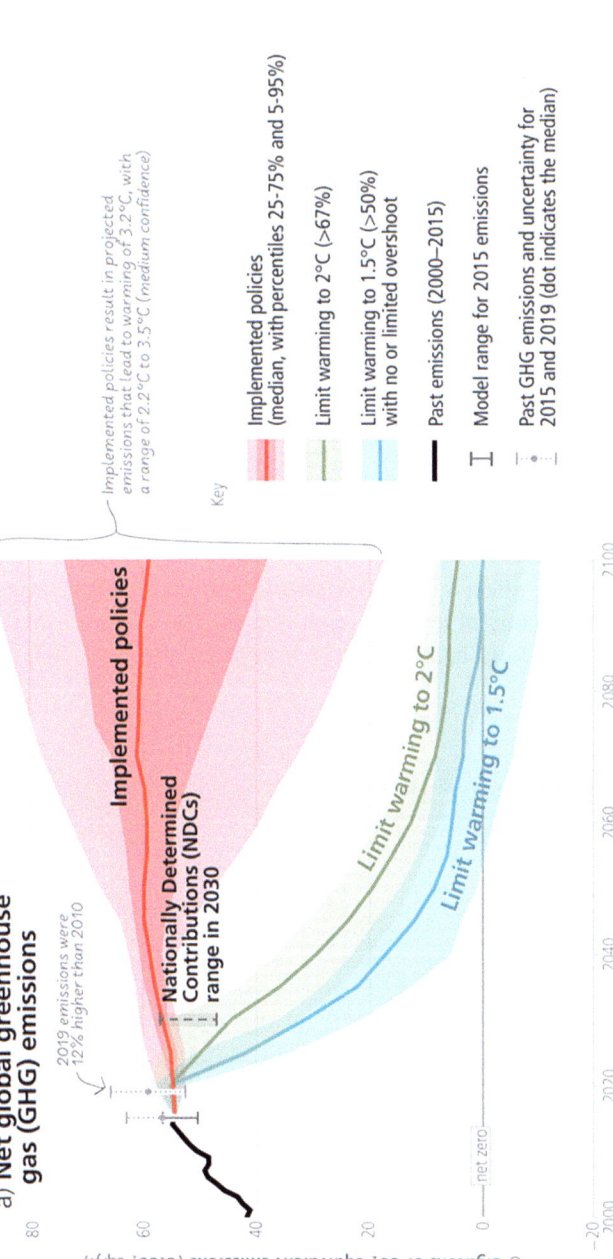

Fig. 2.1 Temperature scenarios from IPCC. Limiting warming between 1.5 °C and 2.0 °C involves rapid, deep, and in most cases immediate Greenhouse Gas Emission (GHG) reductions. Net zero CO_2 and net zero GHG can be achieved through strong reductions across all sectors. *UN IPCC Report, 2022*

challenge the resilience that nature provides against climate change. All this will result in migration of large groups of people that within a short time will move toward areas still suited for food production and without severe drought. The migration caused by climate change and damage to food systems and nature is expected to cause the largest human migration seen in decades if not centuries.

Politicians and business leaders have been presented with information and warnings from the IPCC annual COP (Conference of Parties) meetings held yearly since 1995, and they have been made aware of human-induced climate change for decades. Politicians, business leaders and other important institutions have ignored these warnings, even though the scenarios presented by the IPCC for years have shown dramatic consequences. Meetings and negotiations have been held and targets have been set but action has been little. It now transpires that the oil industry and the automobile industry have been funding scientists since the 1960s and have been informing industry leaders about climate change caused by fossil fuel burning for half a century without them responding to these warnings.

Scientific Background on the Greenhouse Effect

The greenhouse effect is a well-known and well-documented theory that has existed since the beginning of the eighteenth century, and is illustrated in Fig. 2.2 (Haar, GHG Effect Drawing based on NASA information material, 2021). It describes how

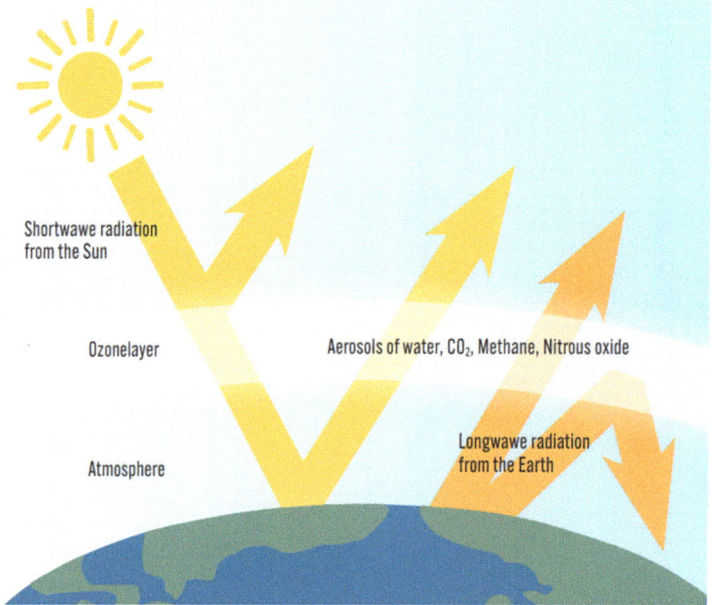

Fig. 2.2 The Greenhouse Gas Effect. Illustration of the Greenhouse Gas Effect. Source: Drawing based on NASA information material

gases and water vapor in the atmosphere withhold sun rays and heat the atmosphere, providing the temperatures that foster life to organisms such as animals, plants, and protozoa (unicellular organisms) on earth. To contest the greenhouse effect itself and the fact that the gases impact the temperatures of the earth is nonsense.

The most important greenhouse gases (GHGs) are water vapor (H_2O), carbon dioxide (CO_2), methane (CH_4), nitrous oxide (N_2O), ozone (O_3), and freons (CFC). Ozone creates a shell around the atmosphere that protects living organisms against some of the most damaging sun rays. Freons (CFC gases) hold a strong Green House effect, and are gases produced from industrial processes, and an increasing source of climate change. These CFC gases are replaceable and easier to limit than the other gases. Unfortunately, it has not been possible to eliminate these CFC emissions globally due to the transfer of industrial production to countries with no legislation on this. Some CFC gasses impact on the ozone layer by creating holes in the protective layer of ozone (O_3) in the outer atmosphere. The hole in the ozone layer above the North Pole has narrowed since the Western world realized this problem and limited the emission of CFC gasses through legislation. Now the emissions are from the production in Asia, and the problem that was almost solved in the Northern Hemisphere is now reappearing above the South Pole. These GHGs originate from burning fossil fuels, industrial processes, deforestation, and agriculture such as animal husbandry and farming of land.

GHG often transforms into CO_2 equivalents ($CO_{2\text{-eq}}$), as a conversion of all types of greenhouse gases.

GHG and pollution are two very different impacts that are often confused. CO_2 is a Green House Gas and contributes to climate change, AND it is a nutrient for plants and algae necessary for life on the planet. Pollution from fossil fuels comes from other hazardous particles, not from CO_2, when refining and combusting the fuels. These particles are proven to cause cancer, allergies, lung diseases, and generally pollute our environment. The burning of heavy diesel and coal is the worst culprit, and that is why the shipping industry has a particularly large challenge.

Green House Gases (GHGs) as CO_2 create climate change but not necessarily pollution. CO_2 is not a pollutant but a nutrient for plant production by photosynthesis.

The Sun Is the Only Source of Energy

The sun provides all the energy available. All renewable energy (RE) is generated directly or indirectly from the sun and is inexhaustible. The wind, the water streams, and the tide all arise from the sun's impact on the atmosphere in combination with the moon. Just like sun rays are a renewable energy source. The various ways of harvesting this energy directly or indirectly are a matter of storage using different technologies. Renewable Energy (RE) technologies provide the most efficient use of this infinite energy source. Fossil fuel is also energy from the sun that has been stored underground as organic matter hundreds of millions of years ago. Fossil fuels are now hard to access, expensive to exploit, and create pollution and climate change. The energy we utilize from uranium also comes from the explosion and formation of our solar system (Big Bang). The rapid and intense GHG emission occurring now creates climate change that is destroying habitats for many species of plants and animals.

CO_2 as a gas makes the planet habitable and has an impact on all the species on earth, including human life. Plants and algae are primary producers and are at the bottom of our food chains and ecosystems. They live from the uptake of carbon from CO_2. Carbon is one of the most abundant elements on earth and is a brick in all organic materials, thus being important for the biosphere and the building block of all life.

The content of GHG (CO_{2e}) in the atmosphere has changed over time due to geological changes and climate has changed several times throughout the history of the planet. The concentration of CO_2 in the atmosphere has been much higher previously without that being a problem. About 200 billion years ago, when the dinosaurs dominated, the CO_2 of the atmosphere was 1300–3500 ppm, compared to now of approx. 415 ppm. The high concentration of CO_2 then generated the high levels of plant biomass necessary to feed the giants. The climate typically changes slowly, also imposing changes on the ecosystems. When humans create climate changes, we challenge our own way of living by causing dramatic changes in the natural ecosystems and food systems at a speed where ecosystems and food systems cannot adapt. It is impossible for ecosystems to adapt to these rapid changes of temperature and climate conditions that the planet is experiencing now.

Many believe that nuclear power is a renewable energy source, and the EU has recently announced it to be a "*sustainable*" energy source. It may be so when it is not based on uranium, but instead based on clean and widely distributed, available elements. No matter how it is viewed, there is only ONE energy source, the sun.

Some people still discuss, and have been discussing for decades, if the rising temperatures are from human activities. There is no doubt in my mind that the reports, warnings, and recommendations coming from the IPCC are scientific, well-documented, and well-founded, as illustrated in Fig. 2.3 (Karl et al., 2009). Scientifically it is well-documented that the rising temperatures over the last 50–100 years were created by human activity. In 1988 the UN established the Intergovernmental Panel on Climate Change (IPCC) with 195 member countries (https://www.ipcc.ch/). The contributors to IPCC are more than 450 lead author scientists, all professors within different subject matters and 2500 scientific experts as reviewing contributors to the reports, including critics of the theory on manmade climate change. The challenge is that we have not listened to the IPCC over the years, and it is only in the latest years that the extent of climate change has really dawned on people in a broad sense and now we experience the consequences in our daily lives.

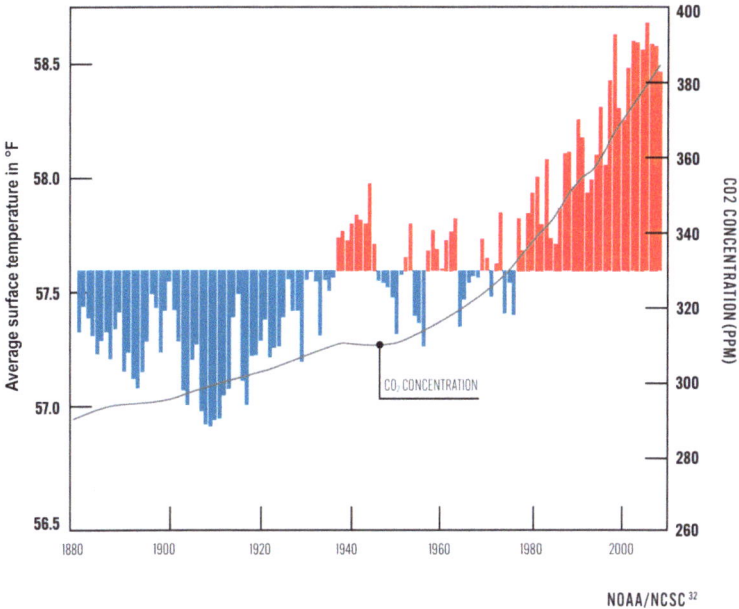

Fig. 2.3 Correlation between GHG and temperature. The correlation between CO_2 and surface temperature, and it is seen that the concentration of CO_2 began to increase drastically after World War 2, when industrialization took off

> There is no doubt that the reports, warnings, and recommendations coming from the IPCC are scientific, well-documented, and well-founded.

Back in 1995–1996, I studied with Professor Svend Jonasson at the University of Copenhagen (KU). He was a member of IPCC and conducted research on feedback mechanisms of greenhouse gasses (GHGs) from the Arctic tundra due to increasing temperatures. Back in the mid-1990s the calculations already predicted so daunting scenarios that they should have required urgent action. Within IPCC researchers then discussed whether to attenuate the scenarios, as they were afraid that they would appear unreliable, and thereby not lead to the action required. The reality now shows that the worst-case scenarios were true and that the climate changes came even sooner than predicted. The increased temperatures, the melting ice, and the extreme weather overtook the IPCC predictions, and the IPCC in 2021 confirmed the man-induced increase in temperature was 1,07 °C and rapidly increasing.

> Carbon dioxide: 415.7 ppm ± 0.2 ppm = 149% of pre-industrial levels.
> Methane: 1908 ± 2 ppb = 262% of pre-industrial levels.
> Nitrous oxide: 334.5 ± 0.1 ppb = 124% of pre-industrial levels.
> Source: UN Climate report, 2022

> Renewable energy is defined as energy supply based on clean and inexhaustible energy as sun, wind, water, and waves.

Consequences of Climate Change

The increasing temperatures will impact on many levels and on all ecosystems as illustrated by the UN in Fig. 2.4.

This story map clearly illustrates the need for change and why the consequences of climate change will be so dramatic that it will change the lives of all people on the planet.

Temperatures are increasing relatively higher toward the poles and melting of ice in the Nordic regions, as Greenland, and from the Antarctic is more rapidly than foreseen. Over the recent years, the increasing temperatures have been accelerating, threatening the habitats of indigenous people, animals, and plant species. For example, the polar bear will soon become distinct due to loss of habitat. The sea level is rising faster than foreseen causing land to disappear and flat country and smaller islands are the first to lose large areas of land. Sea levels have already risen by approximately 28 cm and the IPCC predicts that sea levels will rise by 0.3–1.1 m by the end of this century. This will change many islands and coastal areas and threaten the habitats of most people living here.

Climate change also causes more extreme weather conditions with more rain in some regions and extreme drought in other regions, but also the change in spread of rain over a year is already a problem in many agricultural areas. The drought is causing forest fires, food production shortages, and migration of people, as is already seen all over the planet as in Australia, America, Africa, and Europe. In many countries, farmers are already adapting to changing climates due to summer drought and increased rain. Large areas on all continents are laid fallow because they are not suited for agriculture any longer. Africa is and will experience extensive famine due to drought and is already causing refugee flows within Africa.

The devil's lawyer may ask the questions: What if climate change, despite all science are not human-induces and we start the transition to a Green and Circular Economy free of fossil fuels, and this will not counteract climate changes. Have we then wasted lots of money and limited growth to the detriment of humanity? The answer is, perhaps climate change is not the most important reason to abandon the fossil economy and change our land use. Other reasons are at least as important.

The transition to a Green and Circular Economy has many goals of solving other challenges than climate change. The investments in the green transition are probably cheaper than the costs of doing nothing, and of the costs of wars fought over fossil resources. Many of the wars since WW2 have directly or indirectly been fought on the access to oil. The Green and Circular Economy also contributes to human independence on resources that can be exploited by dictators, kleptocrats, and capitalists without an eye on humanity and the planet. This is clearly demonstrated from the War in Ukraine.

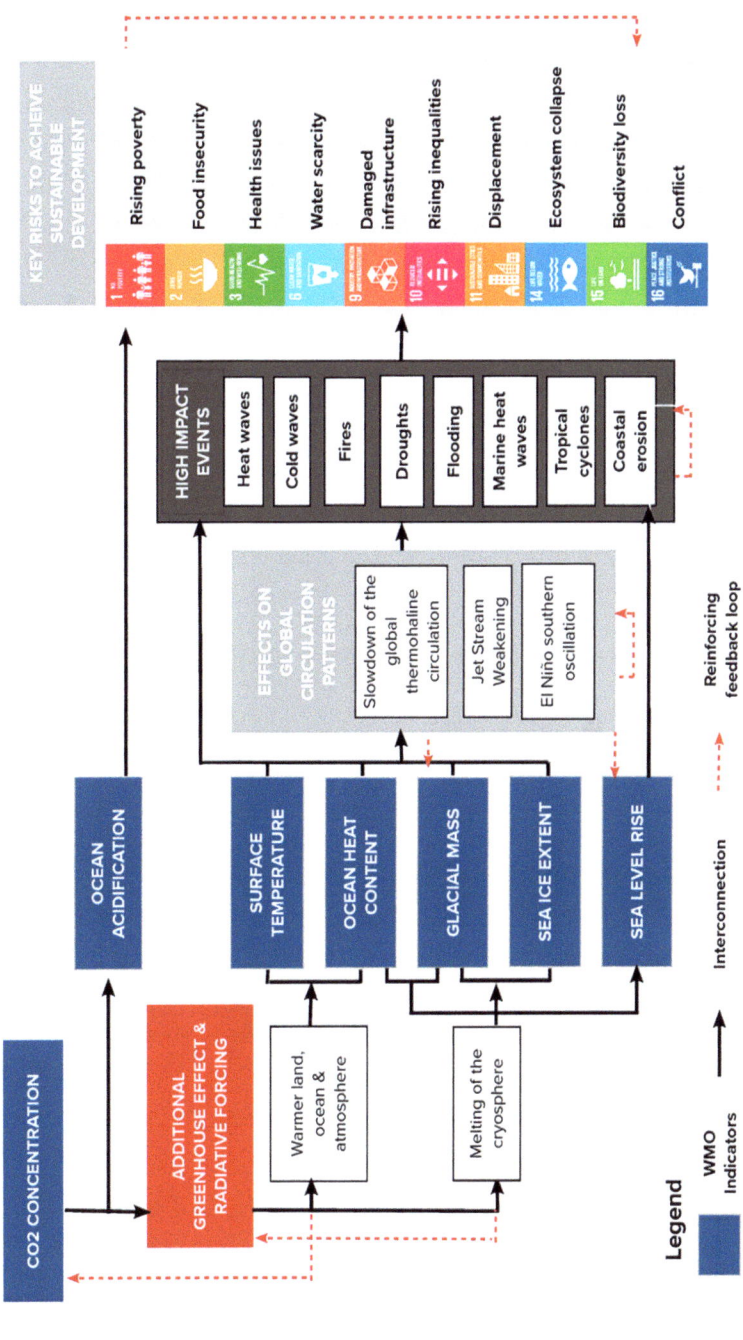

Fig. 2.4 UN story map on consequences of Climate Change. UN story map on impacts of increasing CO_2 levels and increasing temperatures. https://story-maps.arcgis.com/stories/6d9fcb0709f64904aee371eac09afbdf

Other reasons than mitigating climate change for transforming to renewable energy supply:
1. Political stability. A large part of our existence and lives are based on energy resources that are only available in a few places on Earth, creating unstable economies and wars.
2. Eliminating hazardous pollution from particles from the burning of fossils.
3. Supply and price stability.
4. Restoring biodiversity and wild nature as the most important protector of human living.

GHG Emissions

The total global man-induced GHG emissions including land use activities totaled 52.4 Gt CO_{2eq} in 2020, written as 52,400,000,000 tons, which has doubled since mid-1980s and is still increasing by approximately 1–2% per year (JRC, 2019). It has also risen since 2015 illustrated in Fig. 2.5 on country level.

With integrated and ambitious implementation plans from politicians and a lot of companies ready to take off, the goals may be achieved. More action and fewer discussions are needed to reach the 13–30 Gt CO_{2eq} agreed upon. Unfortunately, most effort is spent on discussions rather than action.

GHG emissions are officially reported by the countries and sectors based on emission geography, and not from where the products or services are consumed. This may make good sense, but to understand how best to avoid GHG emissions and create the best and fairest transition, import and export must also be considered in setting goals.

As seen from Fig. 2.5 China, as the largest emitter, with twice the emissions of the USA, is closely followed by the EU27. Together, these three regions account for 50% of global emissions. India, Russia, Japan, and Brazil are responsible for another 17%. So, very few nations and leaders are to agree on changes to mitigate climate change.

Greta Thunberg and her rhetoric split people's opinions, but her data references come from IPCC reports and other scientific reports. She cannot be accused of exaggerating or painting a doomsday picture. She refers solely to published and well-documented scientific knowledge.

> **The Crises we face—the Climate Crisis and Mass Extinction—are so simple that a child can easily understand it**
> —Gretha Thunberg, climate activist.

Unfortunately, the situation is as simple as quoted here and it is just the pursuit of growth, that have made world leaders—both political and company leaders—not to react earlier. There is an urgent need for action, and we have passed the point of no return, as climate change is already happening, and GHG emissions are steadily increasing.

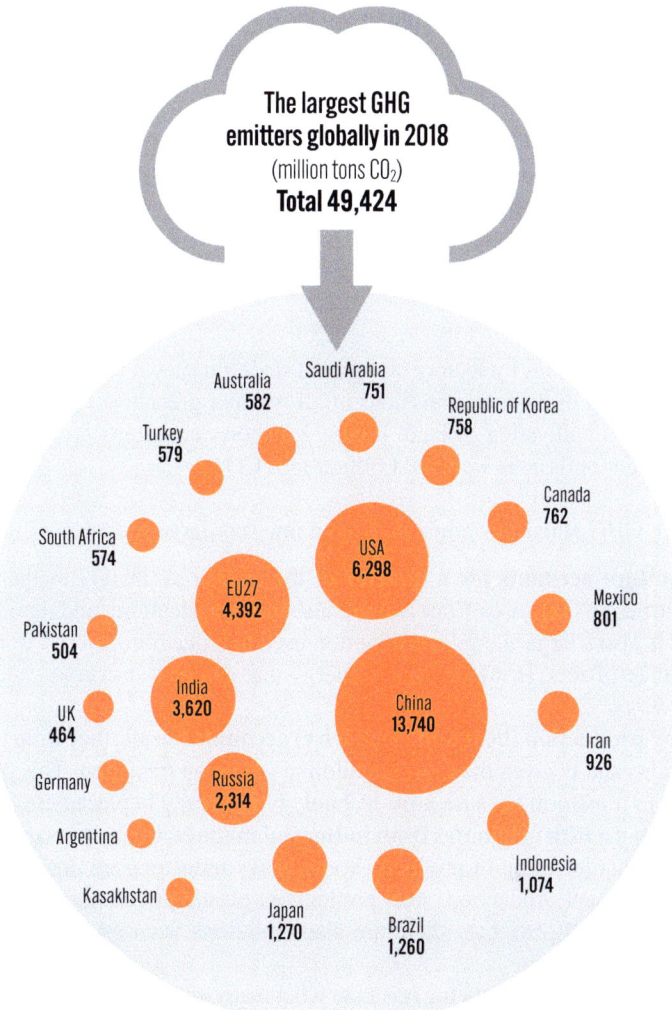

Fig. 2.5 Large GHG emitters by countries. Large GHG emitters by countries. This is shown for 2016—global greenhouse gas emissions were 49.4 billion tons CO2eq. https://www.unep.org/explore-topics/climate-action/what-we-do/climate-action-note/embed.js'data-embed='ghg'

It takes dramatic changes, especially in this part of the world where the emissions are large and where the financial impacts are greatest. All must show responsibility for creating the change needed and each initiative counts. Consumers and every human being are the most important contributors because it is how we act and consume in ways that create changes of scale. When consumers demand changes the companies are listening. The leaders, especially the large corporations hold great power and can create changes in the short term. A lot of the technology that can facilitate the change is available to us. We CANNOT just put money aside for research, as many politicians suggest. We need to implement the solutions we have on hand AND set aside money for future solutions.

> The Earth as a planet will survive—it has tried worse than human existence. The question is whether we want to change the conditions for humans, animals, and plants as drastically as we are doing now and thereby destroying human habitats, cities, and food production.

The GHG emissions by sectors on a global scale are shown in Fig. 2.6 to provide an understanding of the total emissions of all types of greenhouse gases, including emissions from cultivation of land. Figure 2.6 shows emissions for 2016—global greenhouse gas emissions were 49.4 billion tons CO_{2eq}.

As seen GHG emissions originate from a few very large sources:

- **Agriculture accounts for a quarter** of the emissions, mainly methane (CH_4) and nitrous (NOx) gases from animal husbandry, agriculture, and deforestation. There is also a large energy consumption used to fixate nitrogen (N) from the air used as fertilizers. However, this is energy that can be transformed into renewable energy.
- **Energy production (heat and electricity) accounts for another quarter** of the emissions and is given the most attention in the green transition. This part is the easiest to transform because most technology is already in place here.
- **More than a fifth originates from industrial manufacturing processes**, meaning the manufacturing of products. Some heavy industries are dominant in this field, as cement production, steel production, plastic production, and chemical production, and obviously there are also emissions from other manufacturing industries.
- **Transportation accounts for less** than what many expect—namely only 14% of the total emissions. These emissions need special attention due to their rapid increase. We have switched from energy-efficient means of transport, as ship and rail, to very inefficient means of transport on roads over the last 20 years. Air traffic is also growing rapidly and represents an increasing share per kg of goods or per person transported. Especially in the Western world, transport accounts for a large share of emissions. The increasing global prosperity also means a further acceleration of this.

This spread according to sectors is the basis for part II of this book—the Climate Nexus.

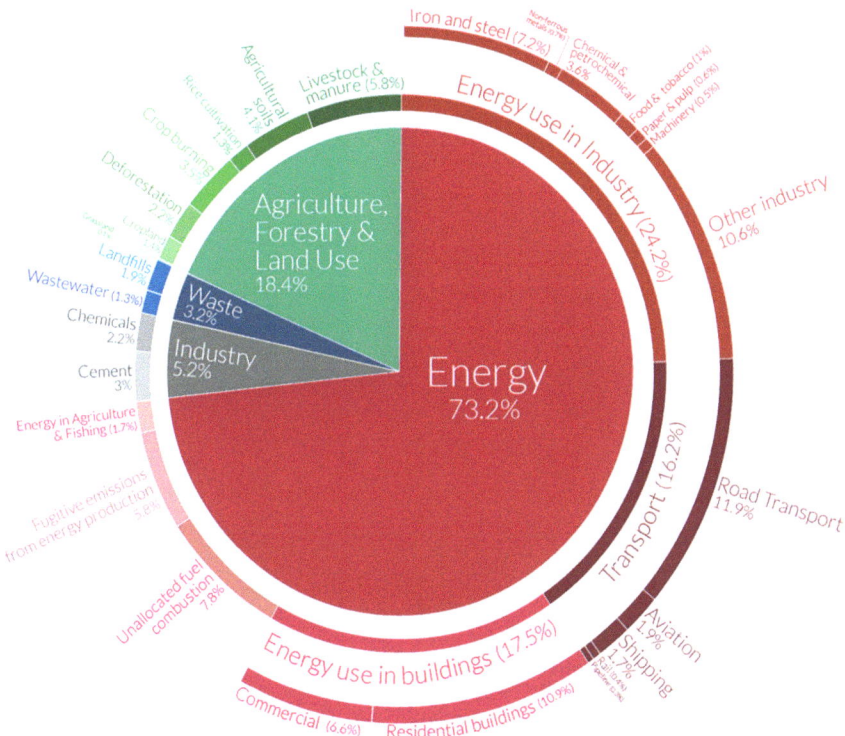

Fig. 2.6 GHG emissions by sector. GHG emissions by sector (2016). https://ourworldindata.org/emissions-by-sector

References

Haar, G. (2021). *GHG Effect Drawing based on NASA information material.*
IPCC. (2022). *IPCC Sixth Assessment Report.* UN.
IPCC. (2023). *UN Climate Report.* UN.
JRC. (2019). *CO2 and GHG emissions in all countries.* europarl.eu.
Karl, et al. (2009). *Global Climate Change Impacts in the United States.* ISBN 978-0-521-14407-0: NOAA/NCDC.

Chapter 3
Land Use and the Spread of Humans

Population growth and the spread of humans over centuries have impacted land use significantly and are the main reason for the state of our planet. As seen from Fig. 3.1 the global population has tripled since 1950. Humans have not changed our demands for virgin resources and consumption despite the increase in population density. UN projects that the human population will hit 10 billion just after 2050 and level out somewhere below 11 billion in 2100. So, still another 2–3 billion people are added to the global population that needs food and decent living standards. The large increase in population comes from two significant events; the invention of the plough some thousand years ago which changed humans into farmers rather than

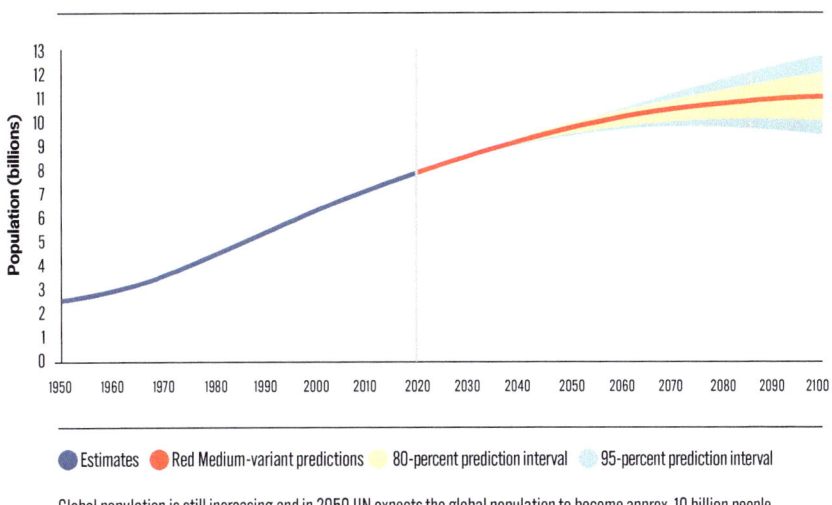

Global population is still increasing and in 2050 UN expects the global population to become approx. 10 billion people.
Source: UN, Department for Economics and Social Affairs. Data Booklet on World Population Prospects 2019 (4).

Fig. 3.1 World population. Global population is still growing and is expected to end at 9.5 and 11.5 billion people in this century. In 2050 global population is expected to be 10 billion

© The Author(s), under exclusive license to Springer Nature Switzerland AG 2024
G. Haar, *The Great Transition to a Green and Circular Economy*, https://doi.org/10.1007/978-3-031-49658-5_3

collectors/hunters, and the discovery of penicillin in 1928, that was put into medicine during WW2.

Population growth has declined or evened out in the old, industrialized world and the population growth has mainly been in the developing regions in the last decades. The incease of the middle class has caused a huge drag on resource and on the land use. The current estimate is that globally 3.2 billion people out of a population of 8 billion people belong to the middle classes, of which 1–1.5 billion live in Europe, North America, and Japan. The middle class has almost doubled in the last 10 years. OECD expects the middle class to increase to a total 4 billion people by 2030. This means an enormous draw on the planetary resources, even though it results in less inequality. Tables 3.1 and 3.2 show numbers on the development of middle class.

Interesting is that between 50 and 60% of global middle class spending will be in Asia Pacific by 2030, and North America and Europe will account for less than half. Still leaving South and Central America, Africa, and Middle East behind. These numbers might be a little pessimistic on Africa where rapid development is seen now.

Table 3.1 Development of Global Middle Class. Development of number (000) and share (%) of the global middle class in regions from 2009 to 2030. OECD data and calculations

	2009		2020		2030	
North America	338	18%	333	10%	322	7%
Europe	664	36%	703	22%	680	14%
Central and South America	181	10%	251	8%	313	6%
Asia and Pacific	525	28%	1740	53%	3228	66%
Sub-Saharan Africa	32	2%	57	2%	107	2%
Middle East and North Africa	105	6%	165	5%	234	5%
Global	1845	100%	3249	100%	4884	100%

Table 3.2 Development of Global Middle Class consumption. Development of spending (billion US$) of the global middle class in regions from 2009 to 2030

	2009		2020		2030	
North America	5602	26%	5863	17%	5837	10%
Europe	8138	39%	10,301	29%	11,337	20%
Central and South America	1534	7%	2315	7%	3117	6%
Asia and Pacific's	4952	23%	14,798	42%	32,596	59%
Sub-Saharan Africa	256	1%	448	1%	827	1%
Middle East and North Africa	796	4%	1321	4%	1966	4%
Global	21,278	100%	35,046	100%	55,680	100%

Deforestation

Population growth and especially our food production has caused deforestation to a very large extent. Deforestation started hundreds of years ago in Europe and has spread to other continents over the last hundred years. Northern Europe was fully covered by forest some hundred years ago. The development of population and deforestation is illustrated in Fig. 3.2.

It is important to understand that forests and natural environments hold enormous amounts of carbon in the biosphere that are released when the land is put into agriculture. The release of carbon comes from the above-ground biomass of trees, scrubs, etc., and a large release of carbon below ground due to composting of the dead organic matter that is in the soils and is not decomposed in the natural habitats. This dead organic matter (DOM) makes fertile soils when decomposed after deforestation and provide nutrients for the crops. After a few years, the soils will become depleted because the organic matter has decomposed and has released large amounts of carbon (CO_2 and CH_4) into the atmosphere. This is why raising forests is important to capture carbon and storage (CCS) and adds to biodiversity if managed correctly. Forest does not equal biodiversity as seen, for example, in Finland, where less than 15% of their nature is protected and biodiverse, despite the huge areas of forests. These forests are part of an industry producing raw materials for wood and paper and are large monocultures of trees.

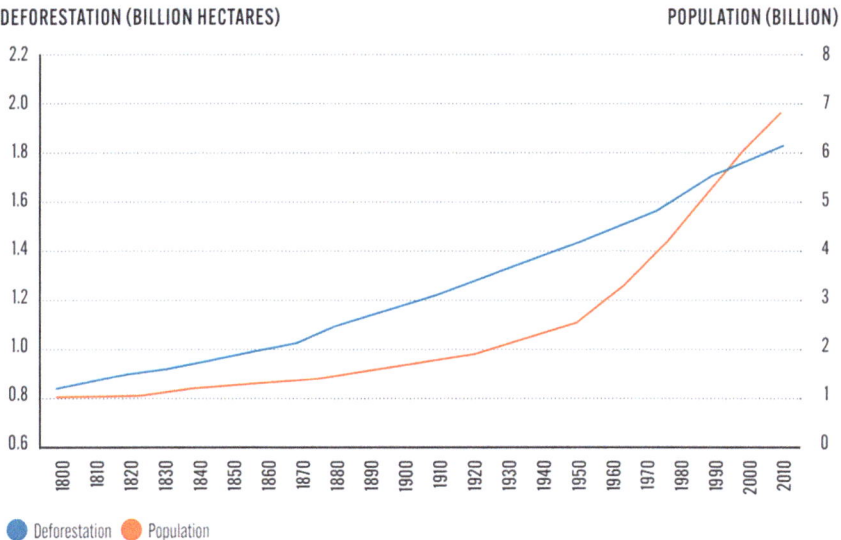

Fig. 3.2 Deforestation correlated to population. Estimated deforestation over time. Sources: Williams, 2002; FAO, 2010b; UN, 1999. Estimated deforestation, by type of forest and time period. Chart from UN FAO's State for the World's Forests (2012)

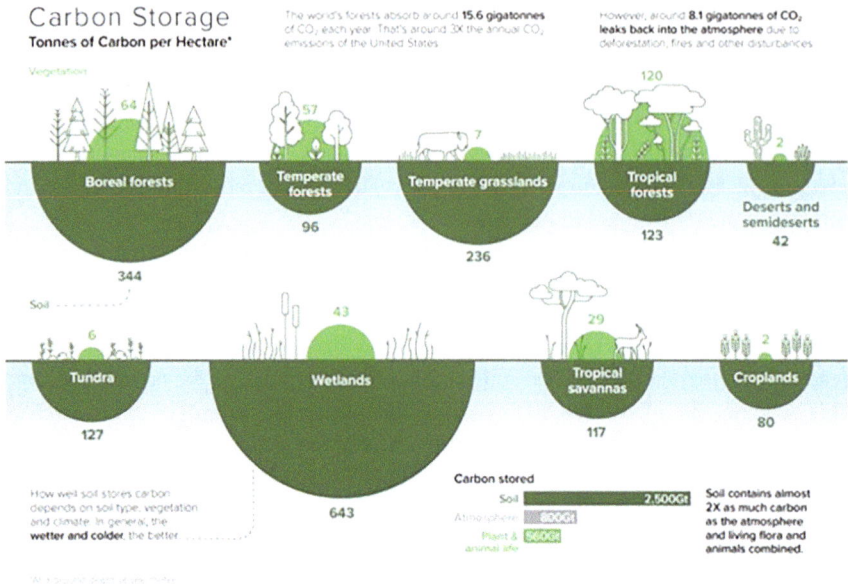

Fig. 3.3 Carbon storage in different natural biotopes. Illustration of the storage of carbon above and below ground of different natural biotopes and cropland. The forests absorb around 15.6 gigatons of CO_2 each year. That is around 3×X the annual CO_2 emissions of the USA. However, around 8.1 gigatons of CO_2 leak back into the atmosphere due to deforestation, fires, and other disturbances. Sources: IPCC, NASA

Figure 3.3 shows the carbon sequestered in various types of nature (biotopes)—above and below ground. The illustration gives a clear picture of the development the last centuries from natural and wild nature to crop land. Humans have developed agriculture by clearing of forests transforming it into agricultural land use. Thereby releasing huge amounts of stored carbon to the athmosphere, especially underground carbon.

Interesting are the large fractions of carbon contained in wetlands and boreal forests, being some of the natural ecosystems that has been eradicated and cultivated into agricultural land, especially in Europe but all around the planet. This illustrates how draining and deforestation in the very populated regions of the planet, as Europe and Asia, have caused the release of enormous amounts of carbon into the atmosphere. A large part of the logs has been used for construction, but still, lots and lots of the biomass above and below ground has been burned, decomposed in the soil layers, and composted releasing enormous amounts of GHG (CO_2 and CH_4).

Reforestation and creating regenerative ecosystems are so important in countering climate change and recapturing some of the released carbon from the atmosphere and into the biosphere again. Forests and wetlands are the most efficient carbon capture and storage systems also necessary for regenerating biodiversity and

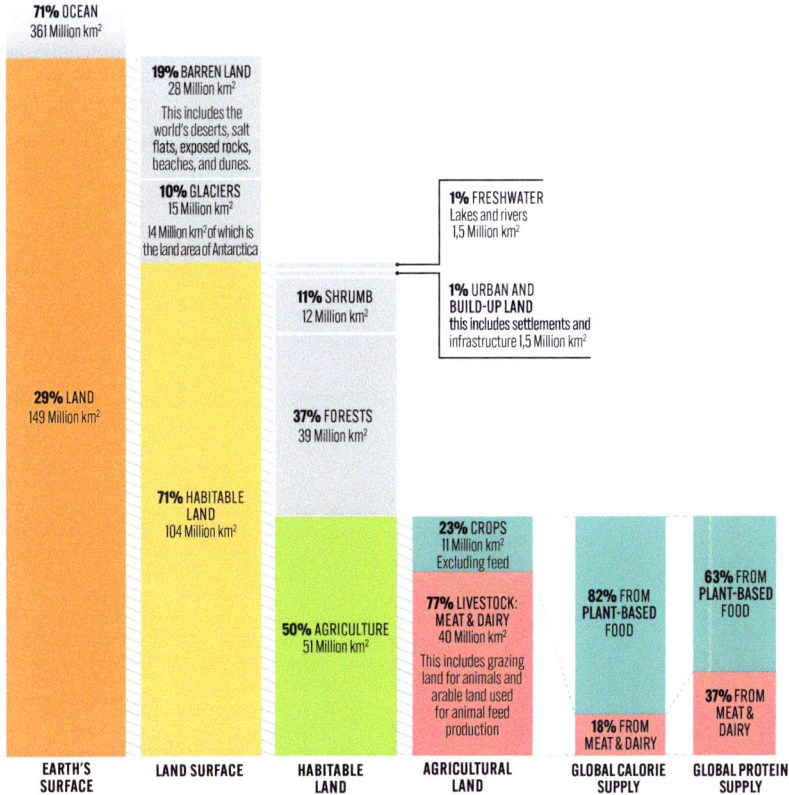

Fig. 3.4 Global Land Use. The illustration shows the share of the habitable land is used for farming and the share is used for meat production compared to the covered calorie and protein uptake. *UN FAO by Our World in Data and authors Ritchie and Roser (2019)*

creating resilience toward climate change and natural disasters. The reestablishment of wetlands is the highest priority of creating biodiversity and protecting against GHG emissions in the global agreements made on biodiversity targets in the UN and EU. Mangroves are an example of a natural habitat that withholds large amounts of carbon and is very biodiverse.

Figure 3.4 shows how the Earth's surface is dominated by humans and how large a share of land is used for agriculture, forests, and shrubland. It also shows that the main part of the agricultural land is used for meat production (71%) even if only 18% of the human protein supply and 37% of the energy supply comes from meat. Meat production is a very resource-inefficient way of producing food with large impacts on land use, wild nature, and climate. This means that if the entire global population were to eat meat at the same level as the high-consumers, it would require even more land to be taken into cultivation. If instead we limit our meat consumption to a healthier level including a low meat intake, we can easily feed the entire population and restore a biodiverse nature—even with a global population of 10–11 billion people. Most of the global food supply is plant based, as seen in Fig. 3.4. The human physiological need for animal inputs (proteins and fatty acids) is very limited.

As seen from Fig. 3.4 most of the Earth's surface is oceans and not land, despite that, we call this planet Earth. The pillars in the diagram nicely illustrate the potential for changing how we manage the habitable landscape to leave space for much more wild nature and biodiversity.

References

UN FAO's State for the World's Forests. (2012). https://www.fao.org/4/i3010e/i3010e.pdf. ISBN 978-92-5-107292-9.

Chapter 4
Resource Scarcity

The old, industrialized economies are experiencing low or no BNP growth. The markets are already experiencing a scarcity of resources, and the struggle for resources has begun. Europe and North America are challenged to secure the living standards of the people, due to inflation, fluctuating employment rates, and dependencies on global product value chains. Europe and the USA are main importers of goods from Asia with China and India being the main exporters of products. Our products once were produced locally and now most of the products we consume are manufactured in Asia and the value chain of our consumption has become difficult to control.

Environmental challenges from manufacturing and landfills have caused environmental disasters. This was followed by legislation in the industrialized countries and then production was moved to countries without regulation on environmental and social impacts. With the EU Green Deal, environmental and social legislations will be covering the full value chain including scope 3, and ensure access to resources and materials locally, and control of minimizing the environmental and social impacts.

Companies increasingly look for a safe and stable supply of raw materials at foreseeable prices. The Corona lockdown demonstrated how vulnerable the supply chains are. It creates great uncertainty when companies are not able to predict from where, when, and at which prices they can purchase their raw materials and input products. Thus, new local markets for recycled materials are welcomed. This will ensure that companies invest in product development and new production set-ups based on reuse and recycled materials. New material loops of reused and recycled materials will shorten the value chains, increase local production, and enhance new business opportunities.

The large fluctuations in energy prices have resulted in economic instability, fluctuating stock prices, and the collapse of nations in recent times and over the last century. Companies' dependency on energy supply and energy prices is a good example of the instability that is now also created by uncertain raw material supply

© The Author(s), under exclusive license to Springer Nature Switzerland AG 2024
G. Haar, *The Great Transition to a Green and Circular Economy*, https://doi.org/10.1007/978-3-031-49658-5_4

and increasing prices on materials that companies are dependent on. Now this unstable situation applies to almost all resources and raw materials that companies depend on.

Industrial and Urban Development

The industrial revolution starting from the middle of the nineteenth century really took off after WW1 and all through the twentieth century. Especially the Western world has developed from an agricultural, rural society to an industrial society. Here access to resources was no longer central. All through this period of more than 100 years, there was almost no scarcity of material resources. The streamlining and increasing productivity of the workforce and the technological development created efficient access and processing of virgin resources. Mechanization and automatization of industrial production along with agricultural industrialization were important in the development of the new urban societies. Large monocultures and chemical fertilizers were introduced to the agriculture followed by machinery instead of manpower as an important start of the urbanization. Goods and food could be produced more cheaply and create growth in the industrialized world.

The industrial moderation in the last decades has been on developing efficiency in workforce and work processes and has resulted in a very inefficient way of utilizing raw materials and our materials resources by the creation of waste.

Figure 4.1 shows the Commodity Price Index (CPI) over the last 120 years (Geronimi et al., 2017). Prices on raw materials have always fluctuated

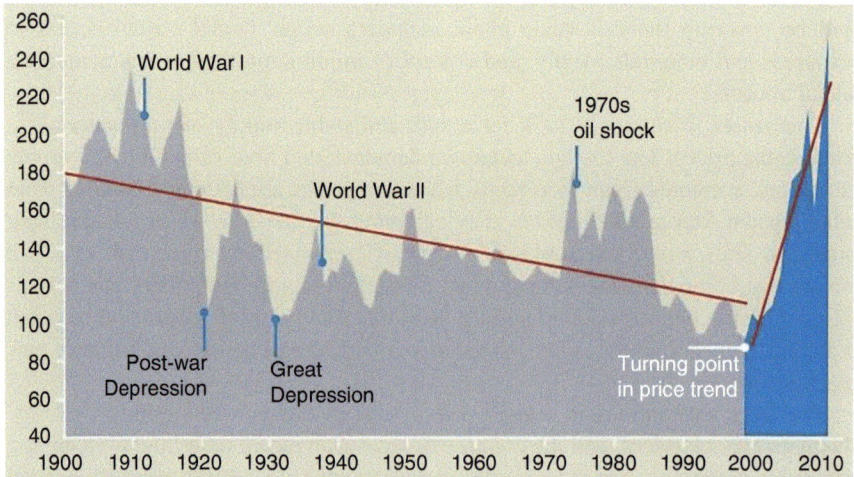

Fig. 4.1 Commodity Price Index (CPI). McKinsey commodity price index (years 1999–2001 = 100), based on the arithmetic average of four commodity subindices: food, non-food agricultural items, metals, and energy (Modified by Ellen Macarthur Foundation) based on the OECD price index (Geronimi et al., 2017)

over years, as those for oil, gas, and coal. These fluctuations are caused by creation of the marketplace rather than by the access to resources and products. The markets are strongly influenced by political instability or new business models causing artificial fluctuations. Prices in a situation of excess resources were not linked to access to raw materials but driven by how products were put on the market. This is called a demand-driven price-setting of goods and this is how prices have been determined in the previous century (1900–2000). The world wars also prevented a free distribution of goods and resources, and materials were allocated to warfare. As seen from the curve wars typically mean increased economic activity and increasing prices. Whereas before and after a war there is an economic recession followed by a fall in prices.

Since the millennium (2000), prices of virgin raw materials and primary food products (not including energy) have increased markedly to a level close to the pre-WWI prices. This recent increase of prices of prices made up with the price declines through the last century due to industrialization, technological development, and intensive agriculture and land mining. The curve, in Fig. 4.1 shows a decline during the financial crisis as of 2010 and then prices have again risen to levels close to or higher than pre-industrialization. The onset of the corona crisis again affected consumption as well as commodity prices. Now prices for some of the crucial raw materials and metals have gone up by several hundred % in a matter of few years. As a new situation, these increases are due to the massively rising demand for raw materials to produce goods for the increasing population, and a significant growth of the middle classes. This is a supply-driven pricing of goods and raw materials, as the opposite of the demand-driven pricing that was dominating the CPI through the 1900s.

The prices for some critical raw materials as Silicium, rare soil minerals, and metals have increased several 100% over a few years. Many of these critical metals and rare minerals are important in the technologies that are required for the energy transition toward renewable energy, as well as digital technologies. Then these raw materials will be particularly subject to price increases in the future.

This means that now scarcity of the resources is influencing the Commodity Price Index (CPI) very rapidly. It is resource-owners and the mining industry that set prices, not market demand. Of course, price fluctuations may still occur, as for example, due to pandemics, wars, and other incidents that change the global activity.

Industrialization in a Historic Perspective

Industrialization resulted in the spread of prosperity in the populations, especially in the countries that also developed democracy. The technological development has made it cheaper to find, mine, extract, and transport the raw materials. Prior to this, colonization provided cheap labor and natural resources without demands for prosperity for the population in the colonies. This means that part of the prosperity that was created in Europe was based on raw materials imported from developing countries—particularly Africa and South America. The processing and refining of products and food was done in Europe due to the technological development here.

The European prosperity also comes from the mining of raw materials locally that feed the new industrial products and machinery and the development of the cities. The steel and cement industries have been important for this industrial development and urbanization. Today, the EU is only mining few raw materials as gravel and stone, chalk, oil, gas, and food. Over the latest decade, the EU, including Denmark, has had a considerable export of energy (oil and gas), and the EU is largely self-sufficient in food, with a small export. However, since WW2 food production has been subsidized by an artificial economic redistribution within Europe. This was necessary after the WW2 food shortage. Today, however, this is causing an economically and often also environmentally unhealthy production along with a distortion of the global development. The outcome today is an agricultural production that is often not profitable and that meets with difficulties to compete on the global food market without subsidies.

Later, also plastic has become crucially important for the development of products. Since the turn of the twenty-first century, the consumption of plastic and the production have doubled—both in the EU and in the countries where our products are manufactured. There is an important difference between plastic and metals in a Circular Economy. Originally good recycling systems were created for the metals, while plastic still only is recycled to a very small extent. Plastic increasingly replaces metals and wood in our products, and at the same time the quality of our product deteriorates to achieve declining prices, as the main competitive "*advantage*." The entire linear production of waste started with industrialization. Introduction of plastic as a new and cheap material in the 1950ies compared to metals and biomaterials has had an important stake in the development of the linear economy, and almost no recycling systems for plastic has evolved due to the low prices of plastic and due to need for plastic waste input for the incinerators for energy production. Plastic embedded in products are difficult to maintain and repair, compared to other types of materials.

The industrial development in the Western world has resulted in an unequal distribution of prosperity and wealth globally. Grotesquely, the countries that today hold the largest and most important natural resources are the poorest countries. While the countries that have driven the technological development achieved prosperity. However, it is important to note that a strong steel and coal industry was created early in Europe based on own resources, especially in England and Germany, but also in the Nordics. This contributed to economic growth and the development of closed loops for metals and glass (sand). Today, we need to move to a Circular Economy to protect the prosperity of Europe. As well as fair access to the markets in North America, Europe, China, and Japan for developing countries must be established, and a higher degree of local processing, both in developing countries and in industrialized countries. More on this in the (Haar, Rethink Economics, 2024) by the same author.

The uneven and unfair global economy has been maintained by the kleptocrats and dictators that held the power in regions that are rich in raw materials and still with a large rural population, such as Africa, South America, and partly Asia.

The advanced industrialized countries have had a steady focus on technology, automatization, and digitalization, also known as Industry 4.0. The potential for continuous streamlining of the workforce is exhausted. Up to 80% of the physical products consumed in Europe today are produced in Asia—mainly in China, but also in the surrounding countries. This means that Europe's dependence on Asia and the Asian economies is huge—as is our supply of cheap raw materials and semi-finished products. That again means loss of control over value chains and how companies manufacture. We have also lost control of the security of supply and transparency in whether products are produced in a sustainable and responsible way.

In that way, we have created global value chains in our manufacturing of goods according to the principles of "take-make-waste" regardless of resource consumption, environmental impacts, or the loss of value by throwing away.

> *"Take-Make-Waste"* means that products are designed for a short lifespan and without the possibility or economic incentives to repair, maintain, and recycle. This is also known from the *fast-moving consumer goods* industry that has affected the way we produce many goods nowadays.

An increasing number of products are being designed according to these principles and over the recent 20 years speeded up overconsumption and created huge amounts of waste. Most of these products end up as input for energy production, or more likely in landfills. The quality of what is put on the market is inferior compared to just a few years earlier. Products hold a short lifespan, and the quality of things is declining. This is true for building materials, textiles (clothes), electronics, and cars, but also for many other things in our surroundings.

The Transition to a Circular Economy

Future prosperity in the EU is threatened since it is becoming more expensive to produce, and because we are dependent on resources not harvested in the region. Our economic models demand constant economic growth, and this is challenging not only for the planet but also for our economies in the old, industrialized countries. Now is the time to create a sustainable economy in a situation of resource scarcity where raw materials come from other sources than virgin mining. It turns out to be cheaper to harvest metals and other raw materials from the big landfills than to mine them virginally. This will become even more profitable when we sort our waste into dedicated fractions and separate metals, plastic, organic waste, etc. at the source (households and companies). Thereby the economic incentives are already there for a Circular Economy without producing waste first.

> Today, it is cheaper to extract metals and other raw materials from waste than from mines.

Circular Economy is about reuse of our products and recycle of materials. In that way, companies and society will become independent of the extraction of virgin resources. The planet cannot supply the growing population with virgin resources if we continue to waste to the current extent, especially in the Western World. Extraction of virgin resources affects nature excessively and our prosperity is challenged by this dependency on scarce raw materials.

Recycling of products and reuse of our materials is not just a consumer demand. It has become a necessity, to ensure that companies continue to have access to scarce resources. Most manufacturing companies are experiencing increasing prices of raw materials and struggle with securing the supply of raw materials and components. This is solved by managing our resources more regionally and by the transition to a Circular Economy. The EU's inner green market in 2030 is based on Circular Economy. The transition will take place within the next 5–10 years for the European companies and the European markets. All are clearly described in the EU Green Deal, see Chap. 6.

An example of the need for a Circular Economy is the construction industry. When renovating or demolishing, building materials are discarded that have lasted for more than 70 years and replaced them with materials that have a maximum lifespan of 20 years and only little maintenance potential. The old, high-quality building materials could easily make another 70 more years, given proper maintenance and repair. The same thing is true for textiles, electronic devices, and many other products. Despite the need for the opposite, the potential of repairing, maintaining, and recycling our resources is currently reduced. There is large financial potential in changing this linear economy into a Circular Economy and speedily trying to maintain some of the high-quality materials in loops instead of replacing them with products of lesser quality. Figure 4.2 illustrates the development from a rural resource economy through a linear economy to a Circular Economy necessary to counter the resource scarcity that we have experienced since the new millennium.

There has been an acceleration of consumption and waste production in the Western World, which means that there is great potential for growth in not letting these resources go to waste and finding new ways of consumption.

Circular Economy is not a new way of thinking. The old and existing rural communities are focused on the preservation and recycling of resources. Circular Economy has created prosperity for centuries in most of the world. People from the less developed regions still live in circular manner due to their limited access to resources and consumer products. People living closer to food and goods production are often more aware of how things are produced. A transition to Circular Economy is not Utopia and some still hold the skills. Circular Economy means changing the way companies organize themselves and getting a handle on the global value chains. Circular Economy is a combination of old virtues and new technology to make the transition effectful and scalable. Outsourcing has created global value chains in the production of stuff, which has fed the business model of "take-make-waste."

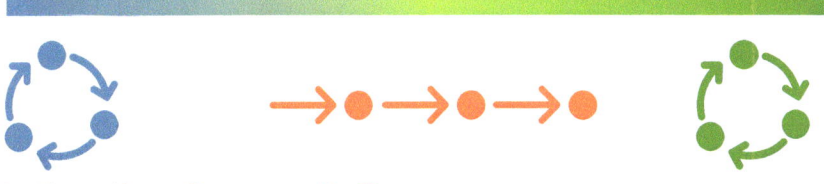

Fig. 4.2 The historic development of the economies. Circular Economy is not a new phenomenon. Originally the economies of societies have been limited by the access to resources. A Circular Economy is a re-establishment of an economy that is limited by the access to resources managed by legislation, traceability, and transparency

In the linear economy, we have lost sight of what we are buying and consuming, even though many consumers want to buy sustainable products. There is a need for transparency and traceability in the manufacturing and marketing of products and foods. Then consumers can make sustainable choices, and companies to make the right decisions and adapt to the future Circular Economy. Consumers and manufacturers must come closer to each other to reach a mutual understanding of demands for products, manufacturing processes, and sustainability.

This means that impacts from products need to be documented in the full value chain and communicated much clearer and claims of sustainability must be based on scientific and comparable data. The prerequisite for creating an overview and thereby credible communication of the global value chains is that validated data about our resources, products, and manufacturing processes are publicly available.

Through the industrial development there was no shortage of raw materials. It has always been a limited share of the world population that has increased their consumption and has been drawing on the resources due to industrial development. Particularly the middle classes are growing fast globally. Growing middle classes means growing demand for food, energy, and consumer products. More and more people are getting access to the same needs of consumption as in the developed world. These are people who move from poverty in Asia, Africa, and South America and into the middle classes. In these regions, there has to a large extent been extreme upper wealth and extreme poverty. The development of a middle class here is an incredible success for global development, human rights, and not least a precondition for democracy in parts of the world that lack this. But it also creates an enormous draw on the planetary resources.

The green transition shall create a sustainable way of living, which means a showdown of the linear business models and the consumption patterns that are based on "take-make-waste."

The whole concept of waste entails the loss of large values, natural resources, biodiversity, and wild nature. That mindset is of the past—if we are to create a sustainable world with prosperity for everyone. Companies need fundamentally to change the way they run their business, and must now contribute to restore nature, ensure access to recycled resources, and genuine sustainable business.

The Circularity Gap Reporting is released every year for the Annual World Economic Forum (WEF). Last year (2022) the report showed that the world is 8.6% circular, which means that 8.6% of our resources are recycled and over 90% of the 100 billion tons of extracted resources are wasted every year. This is an enormous loss of values and resources that we and the planet cannot afford any longer. This also causes environmental disasters, as plastic in the oceans, pollution, and huge landfills in developing countries. Remobilizing these wasted resources is necessary to support consumption in the industrialized countries and the growing populations.

The Gap Report has concluded its recommendations in Fig. 4.3 and gives a good overview of the enormous change that almost any country is facing. Countries have started ordering Circularity Gap Reports on national levels and the Nordics come out below average as expected: Norway 2.4% and Sweden 3.4%. The Danish CG-Report is expected in late summer 2023. The countries with high rates of recycling are typically the low-income and resource-scarce countries, called *build* countries in the analysis. That leaves a big task for the developed countries, called *shift* countries, to take responsibility for their (over)consumption of resources. Here, it is important again to note that the developed countries (*shift*) import their products for consumption and manufacturing of our consumption to a large extent occurs in *grow* countries.

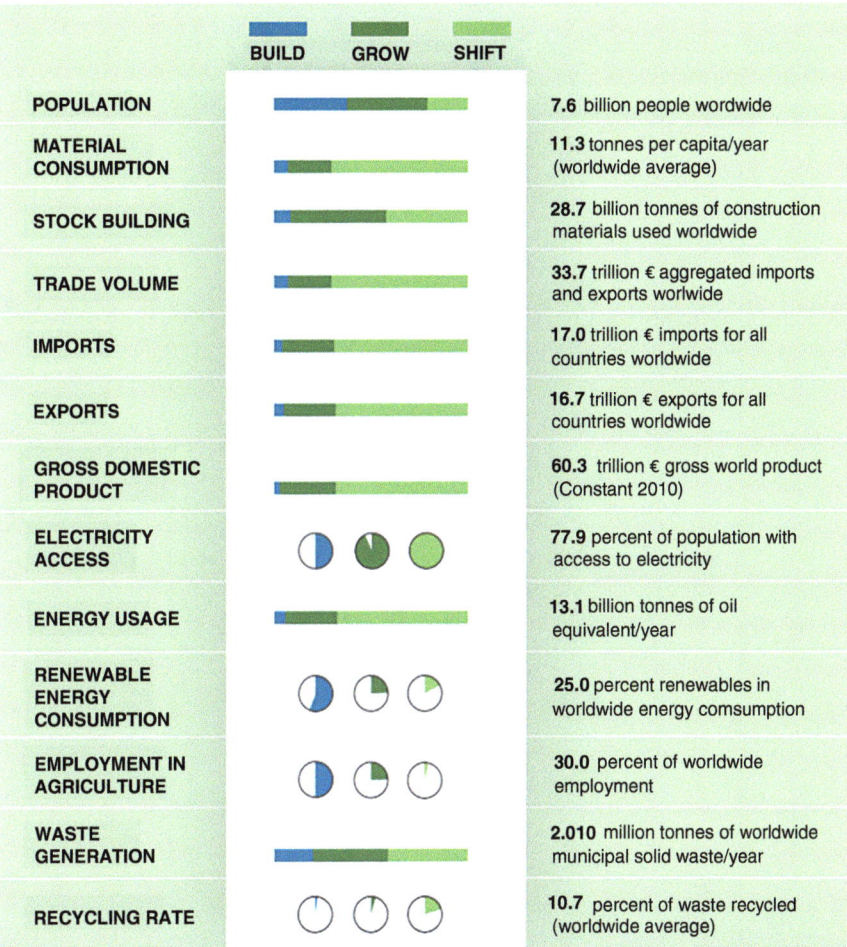

Fig. 4.3 Overview of important physical, social and economic dimensions divided into three types of countries. The data on physical, social, and economic dimensions are included to illustrate the challenges in the three categories of countries: Build, Grow, and Shift explained in the text. *Circularity Gap Report.* **2020**

> **Countries are grouped into three categories: (Source: CGRi)**
>
> **Build Countries** have a low material footprint per capita. As a result, the impact of their economic activities often falls within the regenerative capacity of the planet. On the downside, however, they are struggling to meet all basic needs, not least in relation to HDI indicators such as education and healthcare. Natural capital, rather than human capital, is their dominant source of wealth, which means that the focus is on the extraction and sale of raw materials, while investment in education and skills is insufficient.
>
> **Grow Countries** mostly have already experienced a degree of economic growth and industrialization, which is broadly expected to continue due to a combination of rising standards of living and population increase. As a result, the use of resources in these countries is characterized by fast economic growth and associated material consumption, rapid stock build-up, and an expanding industrial sector (also responding to demand from Shift countries).
>
> **Shift Countries** maintain the highest proportion of services as part of their GDP. Yet, their material consumption is 10 times greater than that of the Build countries. They also produce high volumes of waste, although what they process in-country themselves is usually managed relatively efficiently. With consumption levels exceeding several planetary boundaries, however, the true impact of Shift countries extends far beyond their national borders, with much of the environmental and social costs incurred elsewhere.

References

Geronimi et al., (2017). *Notes on updating price indies and terms of trade for primary commodities*. Research Gate.

Haar, G. (2024). Rethink Economic and Business Models. *Rethink Economics*. : SpringerNature.

Chapter 5
Green Transition and a New Market Situation

We cannot leave it to national states or environmental organizations to create a sustainable planet and common welfare. Fundamentally, new market conditions and legislation is emerging targeting companies and corporations to transform into sustainable business models. The linear business models will not be allowed in the future (EU Green Deal and Strategy for Circular Economy). This means large changes and limitations in which products can be put on the markets. New value chains will occur ensuring that companies get access to raw materials, how they manufacture products, and how products re-enter the loops of reused and recycled materials in the future. This also means a showdown of global value chains and how companies developed their go-to-market strategies. It is the end to produce all that can be sold, in the cheapest and fastest ways.

A way to drive this change is by creating proximity to the value chains and manufacturing processes. Proximity in extraction, proximity in the management of natural resources, proximity in manufacturing the products and the food, as well as closeness to how companies affect the nature closest to us. Proximity to our surroundings and seeing ourselves as a part of nature together with an understanding of how stuff and food are manufactured is the most important step for companies to become genuinely sustainable. Thereby sustainable procurement and active involvement of consumers is an important element in the green transition.

The economic models presented here, are based on classical economic thinking, where the companies operate in a competitive market, although increasing ESG regulation, including regulation on exploitation of nature and the harvesting of virgin resources is expected.

There is still a need for growth and sustainable economic development in many places on Earth. Free trading based on supply and demand as well free movement of capital and people is a proven model for creating economic growth and prosperity among populations. Despite the major challenges in the least developed countries (***build*** and ***grow***), it is here that future economic growth will and should happen. It is important that economic growth is decoupled from resource consumption, especially virgin resource consumption and wild nature. A sustainable impact from people and regenerative utilization of natural resources in close interaction with wild nature, requires stronger regulation of access to the externalities, such as natural resources, the environment, and the climate. This again requires traceability and transparency, so that consumers and businesses can take the best, long-term, and sustainable choices without jeopardizing the planet and the possibility of descent lives of future generations.

> **The virtue called austerity has all but disappeared and with it, the joys of the quiet mind. With the fast development, the businesses have become numerous, the profits greater and the possibilities are unlimited. Everyone wants more.**
> **—Gad, 1913. Danish Author and Playwrighter.**

The Value Chain of Companies Consists of Three Scopes

To understand the elements of a value chain the UN GHG protocol is becoming a general standard for viewing the full value chain. It was originally designed for companies' scopes 1, 2, and 3 and is now a common way to address the scope of the impacts in the value chain. It was also designed to visualize the GHG impacts but is now used to address all impacts in relation to ESG. A detailed description of the UN GHG Protocol is included later in this book in part II and is also included here as the book includes references to scopes 1, 2, and 3.

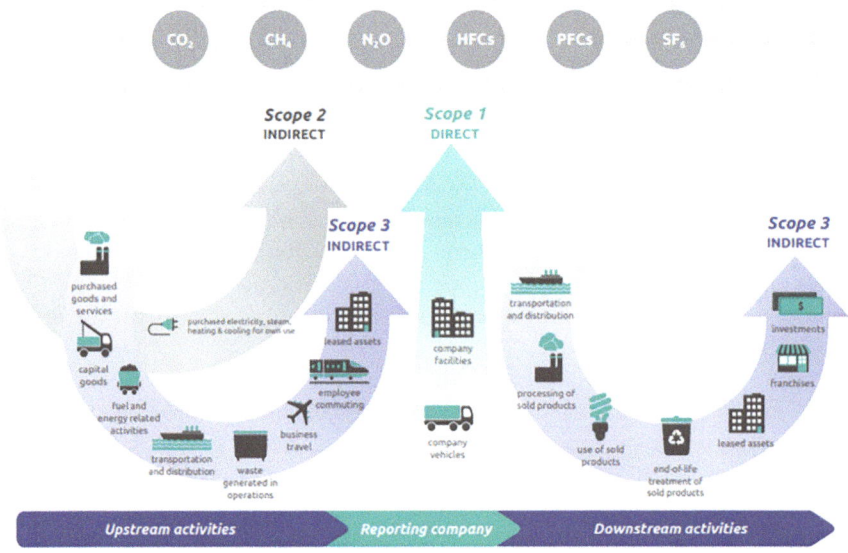

The responsibilities of companies and management are extending to the full value chain—all three scopes—these years, but this is not really reflected in how we perceive the impacts of nations and the responsibilities of national politicians, and how they report on ESG impacts to international bodies.

From CSR to Sustainability and ESG

Historically, sustainability and CSR (Corporate Social responsibility) in companies never built on efforts to secure future markets or on access to critical materials. To many businesses, sustainability still seems like extra work on the side of the real business. Sustainability and CSR are often considered extra costs in the communication department rather than the most significant competitive advantage and the foundation for the strategy.

The transition to a Green and Circular Economy is essential for holding a business in the future. Sustainability and businesses go hand in hand, and companies are central wheels in the great transition. Companies have for many years issued CSR reports and then continued business as usual. The maturity of the companies from CSR and philanthropy toward strategic sustainability are illustrated in Fig. 5.1.

Businesses should have integrated sustainability into their core business long ago since it is both profitable and crucial to market access. Consumers are becoming more aware of the impact companies have on the unequal and unsustainable world. Probably because the CSR approach does not evolve from the core business, but

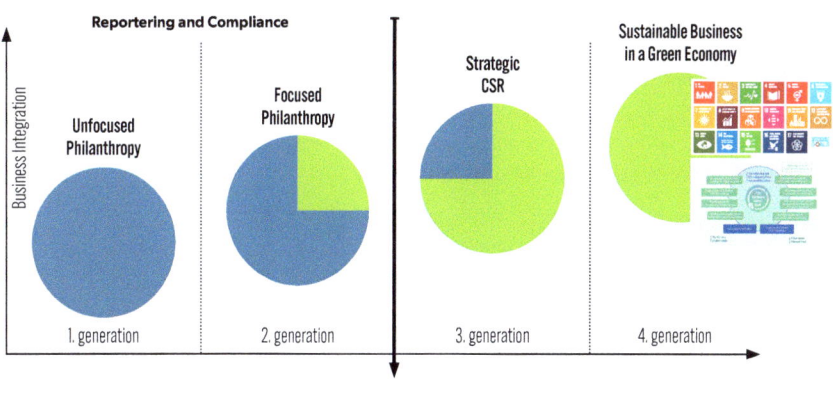

Fig. 5.1 Development from CSR to integrated sustainability. Illustration of the development from Corporate Social Responsibility that companies worked with years ago mainly based on philanthropy to business integrated sustainability based on SDGs and EU Green Deal

from the communication departments. They are responsible for reporting and communication, and not for company strategy. Hence, sustainability and the Green and Circular transition did not become the focus of senior management, nor have they given it the strategic focus that is necessary now and critical for businesses. Maybe also due to the complexity of sustainability and the lack of understanding and competences of the topics contained herein.

Today, the green transition and sustainability are driven by customer demand, supported by increasing levels of legislation. Now a strategic focus on sustainability is a necessity if a company wants to be in the future European market and to some extent also the Northern American markets. The 17 Sustainable Development Goals (SDGs) seem to have set a new agenda for many companies, especially in Europe, but still with little action. Many companies use the SDGs as a communication tool and not as a call for action, somewhat like the CSR reports. Though the SDGs have created a new awareness about the state of the planet and human inequality, also in businesses. In another book by the same author (Haar, Rethink Economics and Business Models for Sustainability, 2024a) the SDGs are described from a Nordic perspective and tools are provided to assist companies in prioritizing the SDGs on a strategic level.

Greenwashing

Greenwashing is widespread in businesses. Companies try to showcase sustainability but still without real change, redesign of products, or new business models. The long and untransparent value chains of our products make it difficult for both consumers and business leaders to create sustainable footprints and sustainable businesses. Instead, a lot of greenwashing has been used in the effort to market almost anything without real change.

The framework of the EU Green Deal, the strategy and action plan on the Circular Economy, and the increasing level of legislation on products have the overall goal to eliminate greenwashing with the new directives on Green Claims being released. This means that companies in the future must:

- Be very accurate and precise when they claim that a product or a service has lower ESG impacts than others.
- All the claims need to be documented in the full value chain (scope 1 + 2 + 3) of a product based on lifecycle assessment.
- Sustainability is a wide word and cannot be used like now. Companies must disclose in detail how they work with ESG and document their work in policies, targets, actions, and resources allocated in their ESG reporting as well as monitor how they handle sustainability issues in and outside their organization (CSRD and SID).

This requires transparency and traceability to create security and documentation of sustainability and to bring greenwashing to an end, enabling consumers to make sustainable choices.

> We need to look at solutions as a whole—only then will we be able to make the right choices. Unfortunately, there is a lot of greenwashing from people that should know better.

A holistic approach is important when saving fossil fuels, adjusting energy supply, and transforming into a Circular Economy. It is not enough only to measure the effects of the individual products or initiatives. We need to look at solutions as a whole—only then will we be able to make the right choices. Unfortunately, there is a lot of greenwashing from people who should know better and there is an army of new advisors, sustainability platforms, brands, and entrepreneurs who have plunged into saving the world with big hearts. Many of the solutions that come out here do not actually have a positive impact on the climate when tested within a scientific framework (e.g., LCA). Therefore, the EU Green Deal and the extensive legislation toward companies on ESG reporting, Circular Economy, Due Diligence, new Ecodesign criteria, and Extended Producer Responsibility (EPR) in the EU is necessary to create a transparent market situation rewarding the companies that show the way to genuine sustainability. Otherwise, we just push the problems ahead of us. See more on EU legislation in Chap. 6.

A good example of sustainable efforts that carry a high cost for the climate, is all the good building materials we are currently throwing away to harvest energy savings. By doing so, the disposal and production of new, linear products end up having more negative climate impact than the energy saved over a longer period. It is NOT a good idea to throw your 50–70–100 year old window out replacing it with one that will last a maximum of 20 years. Then climate change is created, not avoided.

It is important to establish a link between our consumption of products and their climate impact on the one hand and saving energy on the other. Just because a technology delivers clean, green energy, it is not certain that this technology or product is manufactured in a sustainable and circular way.

The Green Transition in a Company Perspective

Many still address the three types of sustainability, the environmental, the social, and the financial sustainability. In this book, sustainability is approached slightly differently, and the financial sustainability is connected to, and conditioned by environmental and social sustainability. To clarify the elements of the environmental part of sustainability and green transition Fig. 5.2 illustrates the green transition from a company perspective and connects it to the three scopes of the UN's Climate Protocol for companies (scopes 1, 2, and 3) indicated in the green box.

This is to make the process of the green transition a little more hands-on for companies, especially SMEs that are not subject to ESG reporting yet.

	UN SCOPES	IMPACTS:
SOCIAL RESPONSIBILITY - Labour Rights, Anti-corruption and Human Rights	1 + 2 + 3	Humans, Chemicals
CLIMATE CHANGE - Lower GHG emissions, Energy Optimization and Renewable Energy	1 + 2	Climate and Environment
RESOURCE MANAGEMENT WITHIN THE COMPANY - Water, Climate Adaptation, Cut-offs, Waste, Chemicals, etc.	1 + 2	Resources, Climate, Environment
NEW CIRCULAR BUSINESS MODELS - Reuse and Recycling in the full Value Chain	3	Resources, Chemicals, Environment and Climate

Fig. 5.2 Green Transition in a company perspective. For companies to get a simple illustration of the green transition mapped to UN three scopes

A) **Social Responsibility** is included here, even if it is more the red transition. Eliminating corruption as well as advancing responsible social production are important elements in creating sustainable global value chains. Abolishing corruption is the single handle that can make the greatest difference to humanity. Corruption and vested interests result in inequality, irresponsible trade, pollution, and exploitation of people. This is also where change is the hardest to achieve because selfishness is part of human survival (see Haar, Rethink Economics and Business Models for Sustainability (2024a)).

B + C) **Climate Change Resource Management to environment impacts** are like traditional optimization projects, and do not require new business models or development of new products. There is more knowledge on energy optimization, renewable energy, chemistry, and business in part III of this book. Energy optimization and optimization of input and cut-off from production are often profitable and common sense. The piggy bank illustrates that companies almost always will save money when saving resources. This type of optimization project is within the company's own physical premises and corresponds to scopes 1 and 2 according to UN Climate Protocol.

D) **New circular business models** where products are redesigned and business models are based on reuse, recycling, and longer life of products, is perhaps the most important part of the green transition. Companies often end up changing their position in the value chain when transforming into Circular Economy. Changing the business models, products and restructuring the whole business is often the outcome of adapting to the new circular market situation.

For almost 10 years, politicians and intellectuals have discussed the concept of Circular Economy. Now it is becoming the target for businesses by law. Circular Economy means big change in all parts of society and is an important part of the green transition. The EU Commission defines the Circular Economy like this:

In a circular economy, the value of products and materials is maintained for as long as possible. Waste and resource use are minimised, and when a product reaches the end of its life, it is used again to create further value. This can bring major economic benefits, contributing to innovation, growth, and job creation.

Here, the definition of Circular Economy is expanded, as:

We are to create financial incitements and business models which ensure the highest possible value utilization of the materials by means of reuse, repair, and recycling with no waste of values or resources and the smallest possible impact on the environment.

Circular Economy is embedded in SDG# 12, Sustainable Consumption and Production. SDG# 12 is probably the most important SDG for businesses in the Western World. Although there is an increased attention on Circular Economy, few people understand it in detail.

Environment, Sustainability, and Circular Economy are quickly becoming business drivers and new competitive parameters. To some industries, the social impacts are important when dependent on global supply chains and manufacturing in other regions. The rise in consumer awareness of responsible labor and human rights is changing the agenda in companies. With the SDGs, it has become clear that the large inequality is supported by long, untransparent, and irresponsible value chains.

We Need Better: Not More

For a long time, sustainability has been about minimizing and doing less bad. Instead of focus on genuine sustainable impact in the full value chain of our products and our daily lives. Now we need better not more.

It requires new ways of thinking if we are to transform into a genuine sustainable living instead of just doing less bad and minimizing our living standards. Driving a car, going on holiday, having a nice hot bath, and eating meat have almost become sinful. The atmosphere of guilt is repulsive to many people and inhibits the large potential in a fundamental green and circular transition. Many have not even tried to find out what is good or bad. It is difficult to find out what really is sustainable, and what is just greenwashing. It is also difficult to understand that doing things completely different may be as environmentally friendly as stop doing it.

There is an increasing awareness on the connection between our (over)consumption and its impact on the planet. More and more shouts for climate action and a sustainable living. Human innovation force and technological development are a part of ensuring a sustainable living and a more sustainable planet. Sustainable consumption requires transparency and traceability of our product and manufacturing processes. The green and circular transition is to do more of the good, instead of less bad. Growth, employment, and new jobs, and export of solutions and technologies are central to the green and circular transitions. To create a sustainable future, we need to close the gap between science within materials, material streams chemistry, technology, and innovation on one side, and business and profit on the other side. It is within this gab that the sustainable future lies. The profitability of the Green and Circular Economy directly correlates with saving the environment.

The linear economy and the present consumption patterns need rethinking. We need to secure access to the resources we depend on—today and in the future. We need to reestablish biological diversity and wild nature with fauna and flora. The natural ecosystems are the basis of human existence to create food security, resilience against climate change and epidemics, and for recreative purposes. Especially within medical and food science the access to genetic variation from biodiversity is important. Stabile ecosystems based on wild nature and biodiversity are crucial to human existence.

Traditionally, people lived in a circular resource economy, as many still do today in a global perspective. Macroeconomist and scientist have the last 10–15 years asked for a more efficient resource management and an elimination of waste production, since the loss of values have become more and more evident in the industrialized part of the world, especially outside the companies. This has among others resulted in a description of the concept Cradle2Cradle as a counterplay to the cradle-to-grave concept, and since many organizations and initiatives have made it their mission to recreate a Circular Economy. Noteworthy initiatives are the Ellen MacArthur Foundation (UK), Nordic Circular Hotspot (Nordic Council), and the ambitious national strategy of the Netherlands. Ellen MacArthur Foundation has illustrated Circular Economy with the well-known butterfly model that gives a good overview of the material streams of products and resources (EMF, 2019). It lacks the connection to the value chains of companies and society but is still good to investigate: See Fig. 5.3.

The Biological Circle consists of the organic materials that degrade over time due to biological processes or due to the physical breakdown of fibers or polymers becoming shorter when recycled numerous times.

The final process in **the Biological Circle** when no value can be extracted from the molecules, will be extracting energy as biogas from the organic matter (low temperature) or incineration (high temperature). In the future traditional incineration at high temperatures will be limited, as technologies for chemical recycling will be developed and very little waste for incineration will be available. **The Technical**

Circle is loops of things and materials that can be kept almost forever if they are recycled in clean loops, as metals and plastic.

Especially in Europe, there is a need for resources to be kept higher up in the waste hierarchy and not using resources for energy production. The circles in the butterfly show different levels on which we are to reuse our products and recycle resources to maintain value and harvest the financial potential. The goal here is to close the loops as close to the center as possible then restraining most value from the products and creating more sustainability for the planet.

As the butterfly model in Fig. 5.3 illustrates the Circular Economy is complex and especially the transition from a linear to a Circular Economy requires in-depth involvement of all stakeholders at all levels of society.

Many business owners or management still do not understand the elements of a transition to a Green and Circular Economy even though it will mean great changes for them. They simply do not understand the impacts it has on their businesses, the new market conditions, or the products. This lack of hands-on knowledge is not so surprising, since only a few companies have transformed their business models for the Green and Circular Economy. In Denmark, profitable, circular business models had political attention as far back as 2010. The Danish Business Authorities wanted to demonstrate local growth and local employment and new types of competitive advantages with profitable, circular business models in Danish companies.

This made Denmark a case study for the Ellen MacArthur Foundation, in which McKinsey mapped the financial potential of a transition to Circular Economy in Denmark (Foundation, 2015). In 2015, it was estimated that a Circular Economy

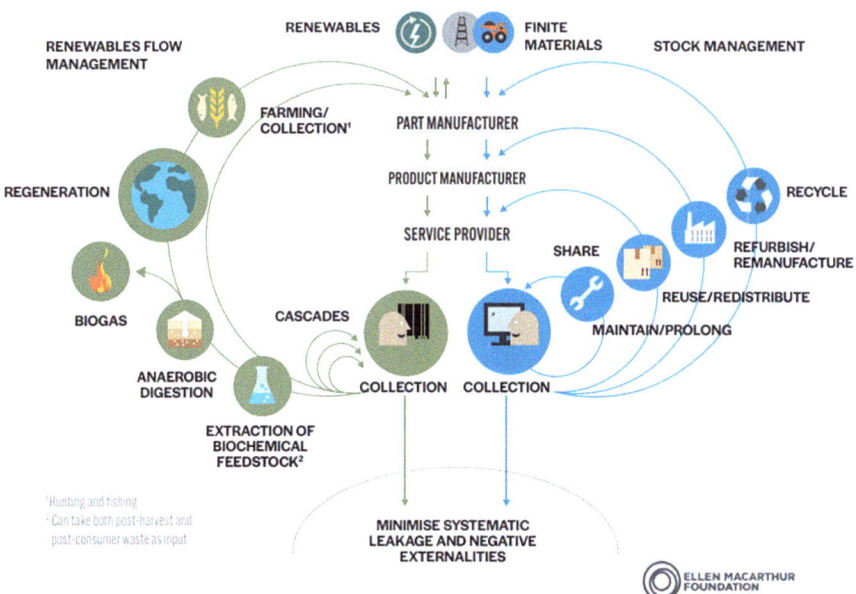

Fig. 5.3 Ellen MacArthur Foundation (EMF) Butterfly Model. Ellen MacArthur Foundation (EMF) Butterfly Model

has a growth potential of 1.5–2.3 billion euros for selected sectors. GDP growth of 0.8–1.4% in 2035 was projected. This potential would probably be bigger today because there now is better understanding of a Green and Circular Economy, and prices on critical materials have gone up since. EU's strategy on a clean and Circular Economy has also set a much broader roadmap toward the Green and Circular Economy. The potential for Europe in implementing the new action plan for Circular Economy is estimated to be 0.5% of GDP, corresponding to 82 billion euros and 700,000 new jobs (Commission, Changing how we produce and consume: New Circular Economy Action Plan shows the way to a climate-neutral, competitive economy of empowered consumers, 2020). Eighty-two billion euros is double the amount of the EU's annual agricultural subsidies and approx. two-third of the EU's total annual budget, so even if 0.5% reads as a small number the potential is huge and Circular Economy can finance development in many European countries because it means more regional production, and this will benefit especially the most challenged European countries. The potential has only increased with the war in Ukraine and the increasing prices of raw materials post-corona.

Some of the companies that transformed their business models back in 2010–2015 are included in a Nordic Case Collection (Haar, Nordic Case Collection, 2024b). These case stories demonstrated the financial potential and competitive advantages of those companies that made use of their first-mover position with new circular products and business models.

Circular Material Loops or Long Value Chains

The Circular Economy is the breeding ground for creating new business models that support take-back systems. In future, the Circular Economy will to a large extent make customers and suppliers one and the same. New customer agreements will evolve because there is a need to secure that materials and products come back to production without detouring as waste first. Access to recycled materials makes companies more resilient to unstable and rising prices for natural resources, and technical materials like rare soil minerals and metals. In that way, countries will achieve regional (EU, North America, Asia) control of the resources recycled in loops. Rather than in linear streams generating waste from long global value chains. It is already cheaper to mine metal resources from landfills than from virgin mines, which confirms the financial potential wasted.

A good and well-known example of a somewhat circular infrastructure is the value chain for wooden fibers used in paper and cardboard and the recycling hereof. The recycling system of today will become more efficient with less environmental impacts when the materials become cleaner without input of harmful chemicals for packaging, printed matter, and other products made from wood fibers. Then a solid material infrastructure (material loops) is available where value can be preserved, and the resources can be utilized to full by all operators in the value chain. A group of three printers in Denmark, Austria, and Switzerland has established an open community: **Print-the-Change** and is working to make the loops of paper and cardboard, environmentally friendly, responsible, and free of harmful chemicals. In that way, a healthy multi-string value chain with both recycling and composting can be established on the back of existing material loops. That will mean maximum utilization of a natural resource where nutrients can be recycled into nature when the fibers are exhausted.

The textile industry include numorous examples of value chains that are broken. In Europe before the 1980s–1990s there was a textile industry that both economically and environmentally was far more sustainable than the current ones. Mostly because production was local and at a smaller scale, and because the products had high quality and long lifespan. People used their clothes up or gave them away instead of wasting it. Prices on textiles today are far too low, the environmental disasters and the social consequences are enormous, and there is an overproduction in the fashion industry where up to 40% of production goes directly to waste without any use first. A large part of the clothes sold are worn on average 6–7 times before they are wasted. Thus, the textile industry produces enormous amounts of waste and only a small part is reused and recycled. The largest part is incinerated or put in a landfill. Going decades back, the textile industry contributed input to the paper industry, since worn-out cotton was used for fibers in papermaking. These two large industries (textile and paper) were important industries in Denmark, and most European countries had their own textile and paper industries. Today especially the fashion industry in the old, industrialized countries, and especially in the Nordics, have created long value chains with production in Asia that are unsustainable, irresponsible, and linear value chains. Which only to a limited extent has created economic and social growth in the producing countries. Today, there is a resurrection of a textile industry in Portugal and some of the Eastern European countries, as well as in Turkey.

Recycling of metals and material loops of metals are of old age. Metals were the cornerstone of the industrialization and very expensive raw materials. Thereby a large portion of the metals harvested from mines are in continuous loops. These material loops must be further developed and spread including rare metals and minerals.

KALK (Denmark) is another example of a company that has created new business and new products by recycling old lime mortar, and by using old technologies to produce new lime mortar for new brick construction. In that way new houses can become breathable, and the lime mortar can be removed from the used brick enabling both the mortar and the brick to be recycled and reused. The new business case with lime mortar is more profitable than with cement mortar, even though lime mortar is more expensive, due to the potential of reuse and recycling.

Companies that have a strategic approach to the green transition will often be able to create a business model that is more profitable due to the competitive advantages of producing environmentally friendly products at the same price or cheaper. This is supported by the increasing demands for traceable and sustainable products and the increasing legislation on extended producer responsibility in the EU market.

Consumer Demands for Sustainable Products

Now, all over the world there is a rising wave of consumers, who are environmentally conscious and socially responsible because it has become obvious that there is a need for change of lifestyle and consumption patterns. They want a lifestyle that makes up with greedy consumption, without relinquishing basic welfare. Often these consumers are not as price sensitive if the products have a well-documented ESG profile. Instead, they are moving away from the constant purchase of new stuff focusing on the impacts of products, their quality and lifespan, and the sustainability profile of the company providing the products or services. There is an increasing focus among young consumers on economic inequality between the producing countries and the consuming countries due to online communication. Especially young people feel closer to their fellows around the globe.

This means that the market conditions for genuinely sustainable products are changing and now companies can price their sustainable product into a more price insensitive market. When sustainable products are produced according to the principles of a Green and Circular Economy, they can in the long run be produced more cost efficient or at the same price as traditional products. The prices of virgin raw materials are increasing, and the supply chain of raw materials is challenged. The transition to circular business models based on reused products and recycled materials holds a large financial potential for companies. Even though it takes investments in new machinery and ecolabels.

Consumers and legislation call for traceability and transparency of the way products are manufactured and consumed, in a documented manner. Then it is possible to make a genuine sustainable decision. It has become opaque to understand the real impacts of a product or a service. Consumers are challenged by understanding how and where our products are produced and to rely on the marketing of sustainable products. Unfortunately, the large jungle of ecolabels is creating confusion rather than insight and consumers are losing trust in ecolabels. EU has introduced the Sustainable Product Initiative, new Ecodesign criteria, and a Digital Product Passport (DPP) with the Product Environmental Footprint (PEF) based on lifecycle analyses to create uniform documentation standards and to euthanize greenwashing. This together with the regulation on Green Claims and the Product Environmental Footprint (PEF) introduced by the EU consumers will be equipped with information to make good sustainable choices.

The sale of reused products is increasing very rapidly due to the demand for responsible consumption and a wish for less "*take-make-waste.*" This industry or new business models are to a large extent driven by small private consumers or SMEs online, and to a large extent a C2C business model. There is a huge potential for manufacturers and corporations to develop business models that promote the reuse of products including repair and maintenance also B2C and B2B. This will be supported by the extensive legislation under the Strategy and Action Plan for Circular Economy as well as the Extended Producer Responsibility (EPR) that is introduced in all product groups.

Statistics on the sale of reused products shows that 8 out of 10 consumers have bought reused products within the last year. In Denmark, every consumer spends €670 yearly on reuse adds up to a large market of €4billion in a country with less than 6 billion people. When asked between 24% and 39% of consumers in selected EU countries bought more used goods online in 2023 compared to 2021 (EU survey on e-commerce shopping behaviour, 2023).

Overproduction and Waste

EU generates 2.151 million tons of waste a year or 4.808 kg per capita (EC, Waste Statistics, 2020). The industries that generate waste in the EU are seen in Fig. 5.4.

The environmental impacts from waste due to overconsumption, overproduction, production to waste and lack of circularity require large changes in the following industries:

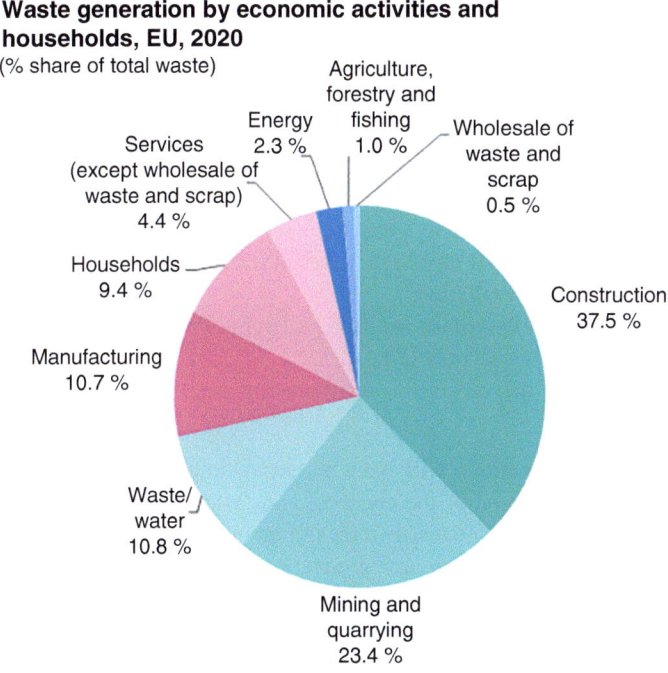

Fig. 5.4 EU waste generation by economic activity. Chart on the waste generation in the EU by sector/economic activity illustrates that construction and mining are the large waste generators. EUROSTAT https://ec.europa.eu/eurostat/statistics-explained/index.php?title=Waste_statistics#Total_waste_generation. Source: Eurostat (online data code: env_wasgen)

- Plastic industry with new design criteria, ban on single-use plastic, ban on plastic exports, and EPR on packing, and high recycling rates.
- Building industry with a large overproduction, low rates of recycling and reuse.
- Textile industry.
- Electronics.

The production of waste is enormous when only 8–9% of all resources are recycled globally. Waste is a huge wastage of raw materials, embedded value, and financial resources in a world of increasing population and demands for prosperity. The linear economy, outsourcing and long, global value chain based on serial production has created overproduction in many industries.

The worst industry in overproduction is the fast-fashion industry and numbers of up to 40% overproduction are reported. Meaning that only a little more than half of the clothes produced end up with consumers. This is to release 6–12 new seasons a year creating a demand for fast and cheap fashion. Much of this overproduction will not even reach the market and is disposed in landfill or burned. This is because the business model for fast fashion is so attractive and production costs so low. It is riskier to run out of clothes on the shelves than to overproduce even to this extent. The fast-fashion industry contributes to the linear economy with lower and lower quality textiles that are less suitable for reuse and recycling. The mixing of fibers has become so extensive that recycling on a fiber level is challenged. Clothes comprise 81% of EU textile consumption and EU released a Strategy on Sustainable and Circular Textiles in March 2022 (EC, Strategy on Sustainable and Circular Textiles, 2022).

Key actions in the EU strategy are:

1. Introducing mandatory ecodesign requirements.
2. Stopping the destruction of unsold or returned textiles.
3. Tackling microplastics pollution.
4. Introducing information requirements and a Digital Product Passport.
5. Green claims for truly sustainable textiles.
6. Extended producer responsibility and boosting reuse and recycling of textile waste.

Implementing these actions will have extensive implications for the fashion industry and will bring some of the cheapest and fastest products to an end. The environmental impacts from the textile industry are on climate, water consumption, pesticide consumption in the primary production of cotton and oil-based fibers, chemical consumption in manufacturing and wild nature, and biodiversity in land use. And of course, on the social impacts in Asia that we all have seen and heard of in the media.

The fashion industry is rated as one of the three most polluting industries and accounts for 10% of global GHG emissions, which is almost as much as all global transport accounts for. The impacts of the fashion industry and their production of 62 million tons of clothes every year are more significant on the other environmental parameters. For example, is almost 50% of all pesticides used globally used in cotton production, and NGOs have estimated that 10,000 liters of water are used to make a pair of jeans and 3000 liters of water to make a T-shirt. In that respect, we need to reuse the clothes and recycle the textiles to a much larger extent.

Quality is not a coincidence. It is always a result of an intelligent effort.
—John Ruskin

The building industry is the largest generator of waste and up to 38% of all waste comes from the building industry and 23.4% from mining and quarrying, which is part of the overall construction industry, as seen in Fig. 5.4. This is due to linear value chains, overproduction but also due to bad management practices. Huge amounts of building materials are wasted due to wrong measures and due to cut-offs on site that are never redirected to be used elsewhere. For example, enormous numbers of insulation bales are thrown out because they have been opened and only a small amount of the insulation has been taken. The same goes for plasterboards. It is estimated that there is an overproduction of building materials that never are installed between 20 and 30%. All this is addressed in the new EU regulation on the building industry (see Chap. 6).

Globally, 57 million tons of electronics are wasted every year. This is due to its short lifespan and poor repairing. All these electronics contain the rare minerals and metals that we are in need of for the green transition. Still, most of the electronics are exported out of EU and wasted in landfills from where some mine of metals and minerals happens. Every year there are incidents of illegal trucks with electronic waste exporting these valuable materials over the borders of the EU creating environmental disasters in Africa and other places. This needs to be stopped and will be so with the strengthening of the EPR on WEEE in the EU that comes out later this year.

Food waste is estimated to be 25–30% of all food produced that is not consumed as food. This is due to long value chains of fruit and vegetables that rot on their way through supermarkets, as well as poor cooling, packing, and storage in the industrialized value chains. Industrialized food is too cheap to pay for responsible food products and enormous amounts of food are wasted. There is a new industry of biowaste emerging that handles food waste for composting and bio-gassing. This is a good idea if no alternatives are available. But the most impact and the most value is created if food waste is avoided, and the fractions generated are used higher in the resource hierarchy for protein, lipid, fiber, etc. production.

It is sad, that the wealth and prosperity created here in Europe have resulted in such an irresponsible management of raw materials and material resources that we now are so dependent on. The Circular Economy is not just a bit of sustainability it is a necessity for providing for people's needs in the future all over the world. This is why legislation is put in place both in the EU, US and around the world to ensure access to resource and stabilize economies and societies.

References

Commission, E. (2020, March 11). Changing how we produce and consume: New Circular Economy Action Plan shows the way to a climate-neutral, competitive economy of empowered consumers. EC, https://ec.europa.eu/commission/presscorner/detail/en/ip_20_420, EU.

EC. (2020). *Waste Statistics.* Eurostat.

EC. (2022). *Strategy on Sustainable and Circular Textiles.* eur-lex.europa.eu/legal-content/EN/TXT/?uri=CELEX%3A522022DC0141: Eur-lex.

EMF. (2019). *EMF based on drawing from Braungart & McDonough. Circular economy system diagram.* https://www.ellenmacarthurfoundation.org/circular-economy/concept/infographic(Online).

Foundation, E. M. (2015). *Potential for Denmark as a circular economy.* https://www.ellenmacarthurfaoundation.org/assetes/downloads/20151113_DenmarkCaseStudy_FINALv02.pdf. EMF

Haar, G. (2024a). Rethink Economics and Business Models. *Rethink Economics.* : SpringerNature.

Haar, G. (2024b). *Nordic Case Collection.* SpringerNature.

Chapter 6
EU Legislation to a Green Economy

It is important to understand the drivers of a new Green and Circular Economy being the extensive legislation and increasing demands from consumers on genuine green products. As well as a strong competitiveness in the market to provide products that serve a greater purpose than just linear waste consumption. The EU legislation and the EU Green Deal are creating significant new market conditions and companies need to transform to have access to the EU markets within the next 10 years.

Even more important is that the businesses need a safe and reliable supply of raw materials and securing of prices for raw materials and inputs. It is a grave uncertainty for a business if it cannot predict from where, when, and at what prices it can purchase raw materials. Therefore, new supply chains of recycled materials will safeguard the innovation and investments made in product development and technological development. New loops of semi-finished goods and recycled materials are important in the circular business models. A good example of the unstable conditions from the uncertain supply and pricing of the goods and raw materials is the large fluctuations of oil prices over the latest decades. With the corona crisis the vulnerability of the long value chains, as well as dependency on resources became clear, for medical equipment but for other life necessities as well. This dependency is spreading into many types of materials and the same vulnerability will occur if we do not find new supply chains based on material loops.

The EU legislation on a transition to a Green and Circular economy is seen as the most ambitious legislation, globally. The USA is moving in this direction, and China is transforming to Renewable Energy and a Circular Economy at a high speed. The challenges here are enormous and the dependency on especially coal is still huge, but the transition here is coming rapidly. Asia, with China as the leader is the main global manufacturing site and responsible social sourcing still has a long way to go, but consumer demand is rising here also.

The EU is continuously working on new legislation that is to support the transition to a Green and Circular Economy under the Green Deal. Many of the directives that are to support the transition to a Circular Economy have already been passed and now Action Plans on creating regenerative ecosystems and a circular

© The Author(s), under exclusive license to Springer Nature Switzerland AG 2024
G. Haar, *The Great Transition to a Green and Circular Economy*,
https://doi.org/10.1007/978-3-031-49658-5_6

bioeconomy are coming out. Over the next 5–8 years, these laws and directives will be implemented in all the member countries. This is to ensure growth, prosperity, and employment in the EU, and thereby making member states less dependent on imported resources. Moreover, the Circular Economy will change the large impacts from our consumption on the climate.

Since the 1980s, the EU has (Brundtland report) worked with scenarios where the declining access to natural resources and raw materials becomes one of the biggest threats to prosperity in many countries. The EU countries no longer have a lot of natural resources available and do not extract large amounts of raw materials from the ground. Earlier, steel production, textile processing, and manufacturing were important industries in Europe, just as there were important metals, as iron ore, copper, silver, and aluminum in European ground. Many of the mines are today closed, and these raw materials are imported into the EU. The EU still has food and wood production also for export but is challenged with the comprehensive exploitation of the land and declining biodiversity.

The President of the USA, Joe Biden, has introduced the Inflation Regulation Act, that actually is a Cliamte Act favoring American production. The ESG considerations in the USA have mainly been driven on a voluntary basis or by market-led responses. Lately, the regulatory landscape is changing in the USA and several initiatives and proposals from the US Securities and Exchange Commission (SEC) are setting national standards on ESG. The Biden administration issued in 2021 an executive order to the federal government to drive assessment, disclosure, and mitigation of climate pollution and climate-related risks in every sector. Various federal organizations have called on financial regulators to focus on capacity building, disclosure, data, and assessment of climate-related issues, as well as pension funds and employee organization are encouraging that asset managers and investment funds are focusing on ESG factors in investments. Sevel demands for disclosure on gender equality and inclusion (D&I) are being put in place in the private sector and financial institutions. It seems as if the focus in the USA is not as broad or as regulated as in the EU. On Circular Economy the US Environmental Protection Agency (EPA) has issued a national recycling strategy as part one of building a Circular Economy. The Strategy is organized by five strategic objectives to create a more resilient and cost-effective national recycling system:

A: Improve Markets for Recycling Commodities.
B: Increase Collection and Improve Materials Management Infrastructure.
C: Reduce Contamination in the Recycled Materials Stream.
D: Enhance Policies to Support Recycling.
E: Standardize Measurement and Increase Data Collection.

The changing focus in the USA towards ESG splits the states and seven southern states have prohibited the states to do business with financial institutions or investment funds that have ESG-oriented investment strategies. So, climate change is still a political battlefield in the USA. Interesting will it be to see what happens after the election and a potentially new president in 2025.

EU seems to have a much more unified and ambitious approach toward ESG and the transition to Green and Circular Economy and has announced that the regulation on ESG disclosure is at the highest standards, thereby enabling corporations to meet future market requirements globally by complying with the EU standards. Now EU has supplemented these anti-global trade initiatives with demands on local production also under the EU Green Deal. The war in Ukraine, post-corona, unstable supply chains and increasing inflation are challenging the economies, the global value chains, and mobility of people and capital. Unemployment and wages are so low that they cannot support a family also in the old, industrialized countries is contributing to the recession that is taking of. The political tendencies toward more local production and more regionally managed economies become real.

The EU Green Deal can also be seen as a part of EU's way of closing the market and taking production back from East Asia, back to EU, especially the eastern European countries or to the countries in the outskirts of EU, as Ukraine and Turkey. We may expect a radical shift in the global economy toward regional economies since they are expected to be more stable and controllable, as well as less dependent on countries and regions with different standards for political control, democracy, and human rights.

The shift to a regional European economy also supports prosperity and employment here, just as it makes the Europeans less dependent on imported resources and products. It is important to understand that our (over)consumption of resources has a significant climate impact, which the Circular Economy counteracts (6). The main framework of the EU Green Deal is illustrated in Fig. 6.1.

The overall goals of the European Green Deal are to:
- No net emissions of greenhouse gases by 2050 with a reduction of net GHG of 50% by 2030.
- Economic growth decoupled from resource use.
- Creating regenerative ecosystems and biodiversity.
- Securing healthy food production and supply.
- No person and no place left behind.

The Benefits of the European Green Deal
The European Green Deal will improve the well-being and health of citizens and future generations by providing:

- Fresh air, clean water, healthy soil, and biodiversity.
- Renovated, energy-efficient buildings.
- Healthy and affordable food.
- More public transport.
- Cleaner energy and cutting-edge clean technological innovation.
- Longer-lasting products that can be repaired, recycled, and re-used.
- Future-proof jobs and skills training for the transition.
- Globally competitive and resilient industry.

The European Green Deal aims to boost the efficient use of resources by moving to a clean, circular economy and stop climate change, revert biodiversity loss, and cut pollution. It outlines investments needed and financing tools available and explains how to ensure a just and inclusive transition. EU has set a goal of a climate-neutral region in 2050. This ambition might move closer due to Russia's invasion of Ukraine

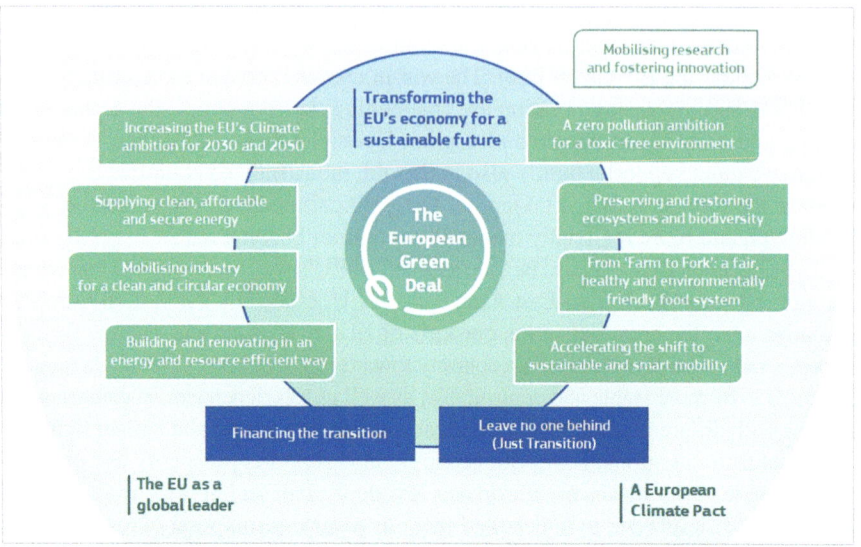

Fig. 6.1 EU Green deal

and the EU's dependency on energy, especially natural gas, from Russia, since some of the large EU countries are discussing climate neutrality already in 2035.

Another ambition is to decouple the European growth from continuous harvest and import of virgin resources and the EU Green Deal has arisen from the same ground as the UN's Sustainable Development Goals (SDG).

Since 2020, EU has speeded up the introduction of directives and regulations and has switched toward more regulation that is applicable directly from EU away from directives that first need national implementation. The extensive regulations on companies and on their products are to create an inner green market in 2030 based on sustainable products and to counter the greenwashing that has been so extensive over the last 20 years. Companies have claimed anything in the name of sustainability without changing the way they produce or take care of the extended value chains. This is over now, and green claims need substantive documentation based on lifecycle analyses. Table 6.1 lists the legislation that companies are meeting now and in the years to come.

All these initiatives together with the Action Plan on Circular Economy, on Circular Bioeconomy (Farm-to-Fork), and on Biodiversity are to establish jointly a coherent policy framework to enhance the development of sustainable goods, services, and sustainable business models as the norm and to transform consumption patterns in a more sustainable direction. They aim to significantly reduce the environmental footprint of products consumed in EU and contribute to the overall policy objective of EU climate neutrality by 2050.

A cornerstone is the new EU taxonomy that will set the standards for SID, SFRD, CSRD, OEF, and PEF. The EU taxonomy is illustrated in Fig. 6.2.

EU defines sustainability as Environmental Impact, Social Impact, and Governance. The challenging aspect of social impacts in the value chains are well known, and international legislation and guidelines have been available for many

Table 6.1 EU legistation on Sustainability and ESG under the umbrella of EU Green Deal

Company legislation:	Year valid
Sustainable Investment Directive (SID) with a new ESG taxonomy on sustainability covering investments from private equity, banks, etc.	2021
Sustainable Finance Disclosure Regulation (SFDR) based on the ESG taxonomy (SID) and screening criteria of investments against asset managers, as private funds, pension funds, and insurance companies.	2021
Corporate Sustainability Reporting Directive (CSRD) is also based on the ESG taxonomy on sustainability. Replacing NFRD and extending the scope of companies. Against other companies that asset managers.	2024–2027 dependent on company size (class)
Corporate Sustainability Due Diligence Directive (CSDDD): EUC rules to drive a responsible and sustainable economy in the extended value chain. It sets rules on corporate behavior and to anchor human rights and environmental considerations in companies' operations and corporate governance. The new rules will ensure that businesses address adverse impacts of their actions, including in their value chains inside and outside Europe. The regulation is put on corporations and directors.	Expected to come out in 2024 with two years to adopt
Environmental performance of products and businesses—substantiating claims as part of the CE Action Plan with close links to a revision of EU Consumer Lax, Sustainable Product Initiative, and Farm-to-Fork.	Proposal period
Organization Environmental Footprint (OEF) or Corporate Environmental Footprint (CEF) is a method recommended to organizations by EU.	N/A
Product regulation:	
Sustainable Product Initiative (SPI) and new Ecodesign Work Plan	Product/industry specific
Extended Producer Responsibility (EPR)	Product/industry specific
Product Environmental Footprint (PEF) based on lifecycle analysis for benchmarking of products to substantiate green claims	Ongoing on product categories

years for companies to commit to and align with, as UN Human Rights Act, OECD Guidelines for Multinational Enterprises, OECD Due Diligence Guidance Responsible Supply Chain of Minerals from Conflict-Affected and High-Risk Areas, The Foreign Corrupt Practice Act, and International Labor Standards, typically monitored under ISO 26000. The game changer here is the six Environmental parameters that cover not only mitigation of climate change but also climate adaption, pollution, circular economy, regenerative ecosystems on land, freshwater streams, and in oceans. The European Sustainability Reporting Standards (ESRS) are described in details by EFRAG, and the framework is illustrated in Fig. 6.2 including the general guidelines that state how companies must monitor and build an internal ESG-governace structure to drive change. According to the EU Action Plan on Financing Sustainable Growth and the Sustainable Investment Directive (SID) these six environmental parameters come with some criteria called the do-no-harm criteria and these criteria are that a company subject to sustainable investments must:

- Contribute positively to a **LEAST ONE** of the six environmental parameters.
- **DO NO significantly HARM** to any of the other environmental parameters.

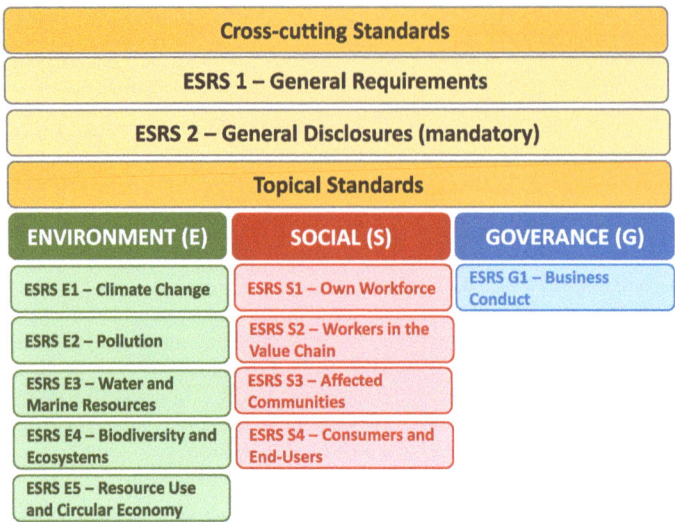

Fig. 6.2 European Sustainability Reporting Standards. The EU Sustainability Reporting Standards (ESRS) is the basis for various reporting legislation as well as the overall topics on product sustainability data

These are extensive criteria that completely change the way companies produces and organizes their value chain, and today no or very few companies can comply with these criteria.

The environmental standards are followed by delegated acts on how the sustainable finance toolbox facilitates access to finance for the transition to a green economy and supporting EU Green Deal. There is established a Platform for Sustainable Finance that accesses and reports on technical screening criteria on the environmental objectives of the EU taxonomy. The platform assesses various industries and their activities and map to the environmental objectives. In October 2022, the EU Commission released a report from the technical working group on methodology and technical screening criteria that is interesting to review (https://finance.ec.europa.eu/system/files/2022-11/221128-sustainable-finance-platform-technical-working-group_en.pdf).

The ESG methodology is well known among corporations in their work with sustainability and requires a company to manage all three pillars within sustainability and it is especially the transition to Circular Economy that affects companies in years to come.

Circular Economy

The Circular Economy is an important pillar in EU's Green Deal with an updated strategy and action plan for Circular Economy (14) were launched. Many European countries are launching and implementing national strategies, regulations, and roadmaps for Circular Economy and the reuse and recycling of products and resources. The Circular Economy has a high priority in the EU and in many of the countries here.

The European Council decided back in 2019 that they wanted a less offensive foreign trade policy still with a strong diplomacy toward China, East Asia, and North America. The economies are becoming less global and more regional, as both the USA, EU, and China demand local production for their markets. China has always had a strong focus on requirements of local production to enter the Chinese market. Less import into EU and more local production require a Circular Economy to meet the goal of disconnecting the economy from the continuous extraction of virgin raw materials.

EU Circular Economy Action Plan (CEAP) includes 54 action—recommendations and legislative actions toward specific industries and the reuse, recycling, and landfilling of materials, and the overall goal is to transform the European economy into a circular economy. The action plan covers policy areas, material flows, and sectors together with cross-cutting measures to support this systemic change through market conditions, innovation, and investments. There are also announced sectorial strategies for plastic, textiles, building industry, and others. The transition to a Circular economy from a legal perspective is illustrated in Fig. 6.3.

This enormous framework operates on many levels, corporate level, product level, material infrastructure, and recycling requirements. These new market conditions caused by all the legislation make one think that the transition will take a very long time. But now EU is releasing directives and regulations at an impressive speed and the ambitions are to create an inner green market by 2030.

Fig. 6.3 The framework for implementing Circular Economy in the EU. Snail that shows the framework for transition to a CE

So, comprehensive EU legislation already exists to transform the European Economy into a Circular Economy and to create an inner green market, making consumers' and companies' choices easier and transparent for them. The implementation of the legislation is taking place in Member States now and until 2030/35. The legislation on extended producer responsibility embedding circular economy will come on all product categories as seen within building materials, packing materials, and electronics now.

On the specific materials EU has released several initiatives in their strategy and action plan for the transformation to a clean and circular Economy (Green Deal): https://environment.ec.europa.eu/strategy/circular-economy-action-plan_en

These strategies are based on the principles of putting an end to waste by addressing the most waste-generating industries first, as the building industry, electronics, textiles, packaging, and plastic. If we can succeed in a transformation to a Circular Economy for these industries and their products, we have created a new basis for growth within EU. Table 6.2 highlights some of the important legislation from EU toward companies that will change the market conditions.

Sustainable Product Package and New Ecodesign Criteria

All product categories are in scope of this regulation on products except food, feed, and some pharmaceuticals. Requirements are also put on some novelties as significant raw materials (e.g., aluminum). Food, feed, and pharmaceuticals are already heavily regulated and will also be covered by the new Strategy for Bioeconomy.

Companies are faced with the big task to learn and understand all the new legislation on the Green and Circular transition in their own industry. This can be done through industry associations or online (see links in Table 6.2). It is important to follow the updates and releases on strategies and actions plans within Circular Economy, Sustainable Product Policies, and Ecodesign criteria.

Ecodesign requirements are followed by performance requirements AND information requirements, all mandatory and supported by the Digital Product Passport. The ecodesign criteria have the objective to increase circularity and environmental sustainability of the products placed on the EU market.

Digital Product Passport

All regulated products in EU must carry a Digital Product Passport (DPP), as all industrial products. See timeline later. DPP is to become one entry point of all product information. DPP is already in place in many products in EU and outside EU, but these are not based on specific standardized systems. The DPP applies a better tracking and tracing of the product throughout the value chain, and the DPP is a tool to create transparency and unlock circularity that will share product information across the entire value chain, including data on raw material extraction, production, recycling, chemical content, etc. Experiences

Table 6.2 Important legislation and strategies from EU on various product groups

Product	EU Commission Initiative	Key	Other legislation initiatives
Built Environment	Strategy for a Sustainable Built Environment (an overall framework)	• **Construction Product Regulation**, including the possible introduction of recycled content requirements. • Promoting measures to improve the durability and adaptability of built assets in line with the circular economy principles for building design and developing digital logbooks for buildings. • Integrate life cycle assessment (LCA) in public procurement and the EU sustainable finance framework and exploring the appropriateness of setting carbon reduction targets and the potential of carbon storage. • Material recovery targets set in EU legislation for construction and demolition waste and its material-specific fractions. • Reduce soil sealing, rehabilitate abandoned or contaminated brownfields, and increase the safe, sustainable, and circular use of excavated soils. https://www.construction-products.eu/publications/green-deal/	**Ecodesign Working Plan** Sustainable Product Initiative **Fit for 55**
Electronics	Circular Electronics Initiative	Regulatory measures for electronics and ICT including mobile phones, tablets, and laptops under the Ecodesign Directive: • Devices are designed for energy efficiency and durability, reparability, upgradability, maintenance, reuse, and recycling. • Priority sector for implementing the "right to repair," including a right to update obsolete software. • Introduction of a common charger. • Improving the collection and treatment of waste electrical and electronic equipment22 including by exploring options for an EU-wide take back scheme to return or sell back old mobile phones, tablets, and chargers. - Restrictions of hazardous substances in electrical and electronic equipment23. TCO measures https://www.europarl.europa.eu/legislative-train/theme-a-european-green-deal/file-circular-electronics	**Ecodesign Working Plan** will set out further details. Printers and consumables such as cartridges will also be covered unless the sector reaches an ambitious voluntary agreement within the next 6 months. **REACH**

(continued)

Table 6.2 (continued)

Product	EU Commission Initiative	Key	Other legislation initiatives
Textiles	EU Strategy for Textiles	• New **sustainable product framework**, including developing ecodesign measures to ensure products are fit for circularity, ensuring the uptake of secondary raw materials, tackling the presence of hazardous chemicals, and empowering business and private consumers to choose sustainable textiles and have easy access to reuse and repair services. • Incentives and support to product-as-service models, circular materials, and production processes, and increasing transparency through international cooperation. • High levels of **separate collection of textile waste**, ensured by 2025. • Sorting, reuse, and recycling of textiles, including through innovation, encouraging industrial applications, and regulatory measures such as extended producer responsibility. https://environment.ec.europa.eu/strategy/textiles-strategy_en	
Packaging	New mandatory regulations, mainly EPR	• Reducing (over)packaging and packaging waste. • Design for reuse and recyclability of packaging, including considering restrictions on the use of some packaging materials for certain applications. • Reducing the complexity of packaging materials, including the number of materials and polymers used. https://environment.ec.europa.eu/topics/waste-and-recycling/packaging-waste_en	EPR as 2025.
Plastic	EU Strategy for Plastics in the Circular Economy	Specific focus on eliminating microplastics: • Restricting intentionally added microplastics. • Labelling, standardization, certification, and regulatory measures on unintentional release of microplastics, including measures to increase the capture of microplastics at all relevant stages of products' lifecycle. https://environment.ec.europa.eu/strategy/plastics-strategy_en	Single Use Plastic Products EPR on packaging

show that one of the largest barriers in the transition to circularity is the lack of information on the products, their origin, their contents, etc. The DPP is a strong support for legislative partners to monitor performance of the products as well as providing the information to consumers and citizens to make sustainable choices.

The DPP are to be decentral kept systems with access to specific data dependent on the industry or the actor in the value chain. The EU regulatory framework is a set

Digital Product Passport 67

of rules to monitor and secure the DPP based on global open standards. Several DPP-like labelling that are in place are, e.g., EPREL (EU energy labelling), SCIP (substances of concern as chemicals, etc.) databases, and EPD (Environmental Product Declaration) on building materials.

The economic entity that places a product on the market (provider) will be responsible for the DPP and will build the databases and the system to comply with DPP regulations. This entity will typically be the manufacturer, or the distributor and they will be responsible for that suppliers outside the EU providing the needed data. Data will be stored decentral at the manufacturers, the distributor, or their trade organization. Data systems, data security, and sanity will be controlled by a central register of regulation on the data ecosystem of DPP.

The data carrier is not regulated upon, meaning that various types of digital carriers can be used, as line code and watermarks. The formats and the security of data will be controlled, but not the databases themselves.

The expected data that needs to be provided in the DPP are:

- Economic operator's name and registered trade name
- Product identifier
- Economic operator's identifier
- Facility identifier
- TARIC code
- Authorized representative
- Description of material, component, or product
- Recycled content
- Substances of concern
- Environmental footprint profile
- Classes of performance
- Technical parameters
- …

The regulatory DPP systems are standards and protocols, and mandates to develop standards to implement regulation on circularity and environmental sustainability rather than on the digital system itself. DPP data/registry will vary by product and loops of materials and will be implemented successively and can be linked to existing DPP-like digital labels. The EPD for building materials, for example, will not replace the DPP unless it will fully be compatible with the legislation here on.

DPP Timeline:
- Political decision EUC and EUP: May 2024.
- Prepare regulation: End of 2024 or primo 2025.
- Best scenario: The first DPP will be on products during 2025. Batteries as the first product category is 2026/27, and a more reasonable timeline for the first DPP on the market is 2026–2027.

World Business Council for Sustainable Development (WBCSD) and BCG have assessed the implications of these new EU Regulations and launched a series of three publications about this topic:

- The overarching report "The EU Digital Product Passport shapes the future of value chains: What it is and how to prepare" summarizes the policy perspective and corporate guidance and illustrates what a potential DPP scenario could look like for actors along the electronics value chain.
- In "Enabling circularity through transparency: Introducing the EU Digital Product Passport," WBCSD and BCG provide details about the regulation and current uncertainties, analyze the different options that could shape the regulation and identify key implications.
- To help companies achieve these benefits, WBCSD and BCG provide clear, actionable steps on how to prepare now as a company in "Navigating uncertainties of the EU Digital Product Passport: How to prepare now as a company."https://www.wbcsd.org/Pathways/Products-and-Materials/Resources/The-EU-Digital-Product-Passport

Company Reporting Regulation (SFDR/CSRD/ESRS)

The capital participants on the market as banks, pension funds, investment funds, insurance companies, and others are the first that are subject to ESG reporting according to the Sustainable Finance Disclosure Regulation (SFDR) as of 2023, and specific screening criteria are set on how to monitor a sustainable investment strategy. This is to drive the change top-down starting with the capital.

> Data is not the goal; data is the mean to drive change.

Right now, companies are very engaged in understanding the new reporting requirements. The reporting requirements from CSRD are phased in from the annual report 2024 through 2027 depending on company size (class), as:

1. **Listed companies** already covered by the Non-Financial Reporting Directive (NFRD) must report as of the financial year 2024 (Annual Report released in 2025).
2. **Other large companies** that meet two of three criteria: turnover > 40 million euros; balance sheet total > 20 million euros; and number of employees >250 FTE must report as of the financial year 2025 (Annual Report released in 2026).
3. **Listed SMEs, small and non-complex credit institutions** and captive insurance companies must report as of the financial year 2026 (Annual Report released in 2027).

4. **Other SMEs** are not yet stated subject to reporting but can be expected to report as of financial year 2027 (Annual Report released in 2028). Many of these will be meet by requirements from financial institutions and customers and are de facto also covered.

Industry codes (NACE) determine the scope of reporting within the EU taxonomy standards set by EFRAG as the ESRS. ESRS are the detailed framework for reporting.

It is very important for companies to note that it is not only a reporting requirement. The directives require change and a systematic reporting on how change is anchored and monitored in the companies. As the general guidelines of the ESRS also require that:

- **Identify the full value chain**—based on standards value industry chains provided by the framework (expected release during 2023/24).
- **Perform a double materiality assessment** to identify ESG impacts in scope 1 + 2 + 3 as well as financial impacts on company business, meaning both potentials and risks. The MA must include a description on:
    - The reasoning on areas prioritized.
    - The reasoning on areas not prioritized.
    - Develop a framework that states how management (executive and non-executive boards, and consultants and advisory boards) is informed and involved.
    - Monitor how sustainability issues raised by the organization or other stakeholders are addressed and communicated to management.
- **Companies must develop a plan** based on the ESG reporting that includes:
    - Policies
    - Targets
    - Actions
    - Resources allocated
    - Stating the above on short term (1–2 y), medium term (3–4 y), and long term (5 + y)

This means that reporting is now a tool to drive change. No matter how massive it seems right now to report and getting hold of all these data on the ESG taxonomy in scope 1 + 2 + 3, reporting will be the least of the work needed. Driving the change means new ways of sourcing, procuring, producing, new products, marketing, and new business models all along the value chains – upstream and downstream. This means a complete change of strategy in all companies in EU entering an inner green market in 2030 based on a green and circular economy.

> The new reporting requirements in EU (CSRD) are massive, but still the least compared to the changes that businesses must drive to meet the legislation

Part II
Climate Nexus: The Nexus Between Climate Neutrality and Sustainability

Part II introduces the Climate Nexus as a concept where climate impacts and actions are linked to holistic solutions to create genuine, long-term sustainable living for humans. Part II introduces the future and how the great transition is formed embedding the main topics, as:

- Energy transition.
- Transition to a Circular Economy,
- Transition to Sustainable Public and Individual Transport.
- Transition to Sustainable Land Use, Agriculture and Healthy Diets.

Chapter 7
Introduction to Climate Nexus

The Climate Nexus is introduced here to illustrate the complexity in mitigating climate change. Many believe that a transition to renewable energy is the solution to climate change. It is far from! We also need to transform our land use, our value chains, means of manufacturing, and transportation infrastructure if we are to stop emissions ensuring a livable and sustainable planet for all. The Climate Nexus is a framework to visualize the great transition to a Green and Circular Economy. The transition requires a broad understanding of sustainability, impacts, science, business, and stakeholder engagement. If we dare to make bold decisions and have a holistic approach to the great transition to a Green and Circular Economy, the mitigation of climate change can be the train to drive change. But first, we need to understand the details of this transition to genuine sustainability.

The green transition affects many industries and sectors to reduce the anthropogenic impact on the climate and on the planet. We need a transition of our way of living for our own benefit to creating a livable planet for all. We must understand how we regenerate nature and biodiversity also to create independence in the extraction of virgin resources from nature. This requires a new Circular Economy because only when we can reuse and recycle nature can be left to regenerate itself. The circular economy is new value chains, new consumption patterns, and new products and will affect consumers and companies to a large extend. Therefore, this book includes a deep dive into the Circular Economy as this is the largest change that many will experience for decades.

Figure 7.1 illustrates and explains within each of the four main areas of the Climate Nexus and describes the challenges, solutions, actions, and effects needed to counter climate change and create a sustainable world at the same time.

The four areas in the Nexus are:

- Energy Transition.
- Transition to a Circular Economy.
- Transition to a sustainable shared, public, and individual transport.
- Transition to sustainable agriculture and healthy diets.

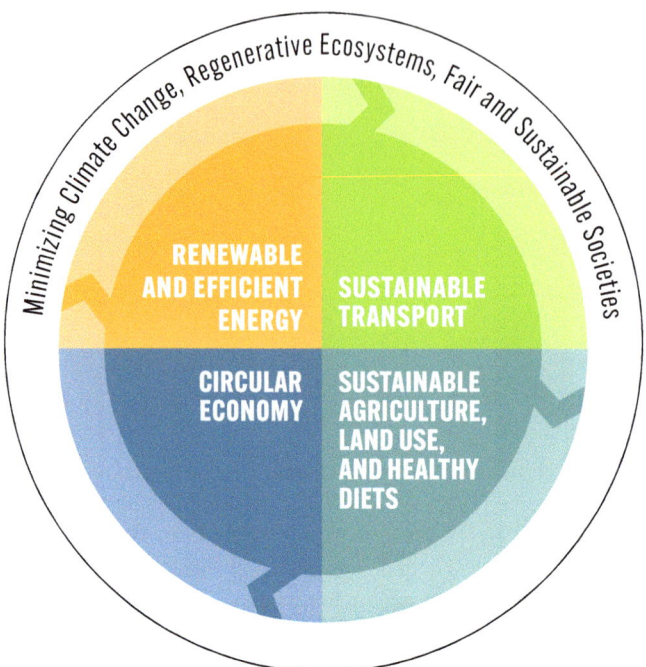

- Political instability from uneven fossil energy source
- Insecure energy supply
- Centralized energy supply
- Particle pollution
- Increasing and unstable fossil energy prices

- Energy inefficient technologies taking over (ICE)
- Particle pollution
- Noise pollution
- Congestion of traffic – time pollution
- Unlivable Cities

- Increasing prices on raw materials
- Resource Scarcity
- Linear business models: Take-make-waste
- Overproduction of building materials, textiles, medicine, etc. never reaching the customer
- Long unstable value chains

- Food waste
- Loss of wild nature
- Loss of biodiversity
- Loss of carbon sources and storage in woods
- Unhealthy diets from industrialized food production
- Over production of meat and inefficient calorie production
- Obesity

Fig. 7.1 Climate Nexus challenging a fair and sustainable world

These areas with a very strong focus on the transition to a Circular Economy are described in the following chapters in this part II in the order presented above.

Chapter 8
Energy Transition

Today, many see climate change as the main challenge created by the consumption and burning of fossil fuels. In addition to climate change, the fossil economy today creates other serious challenges such as:

- Political instability and wars.
- An overwhelming force that influences the distribution of wealth nationally and globally.
- Unstable access to fuels and unstable prices.
- Huge pollution and health problems from particles and noise.
- Inefficient energy consumption.

The links between fossil energy supply, political unrest, and war are often overlooked, though it has become clear again with the war in Ukraine. Many wars since World War 2 have been a direct or indirect fight over access to fossil fuels. In addition to large human costs, this has had considerable economic and political consequences. Fossil energy supply is an unstable supply at volatile prices and always has been. A transition to renewable energy is an important political stabilization that should have had as much attention throughout history the last 100 years, as climate change has today.

After all, democracy is the society model that has given the most people access to education, food, and a decent way of living. **Renewable energy will probably become one of the most democratic tools of our time.** The access to energy is crucial for people to leave poverty and make a decent living. The power over the energy supply is essential in all wars—both as a root cause but also by disconnecting and destroying important energy supply in countries and with troops has a major strategic effect in warfare. That is also why the American military is very far in their transition to renewable energy at their military bases. It makes them less vulnerable, as the supply of energy cannot be as easily sabotaged. A discovered side effect was that they could hear the enemy far sooner, since the noise from the diesel generators allowed intruders to sneak in.

Experience is a very good teacher, but she sends very high bills.

—Minna Antrim

To meet the targets set on climate neutrality it is important to transform the energy supply to renewable energy since it is the most climate friendly and provides the most safe and stable supply. On the way to climate neutrality, there are some considerations that can speed up the transition also with some fossil fuels in the energy system. A very large barrier in the transition to renewable energy supply is the supply of metals, minerals, and microchips to build all these new installations, and all contributions to minimizing GHG emissions need to be considered.

Here, the various energy sources, development of energy prices, and the transition to renewable energy are described. Consumption of energy in a scope 1 + 2 and scope 3 perspective, the industrial production, and the global value chains are also described.

Switching from Fossil Energy Sources

A look at the types of fossil fuels will show where the emissions come from and what challenges they bring (see Fig. 8.1).

The graph clearly shows that since WW2 consumption has increased dramatically, and the dominant sources are oil and coal. The easily accessible oil sources on

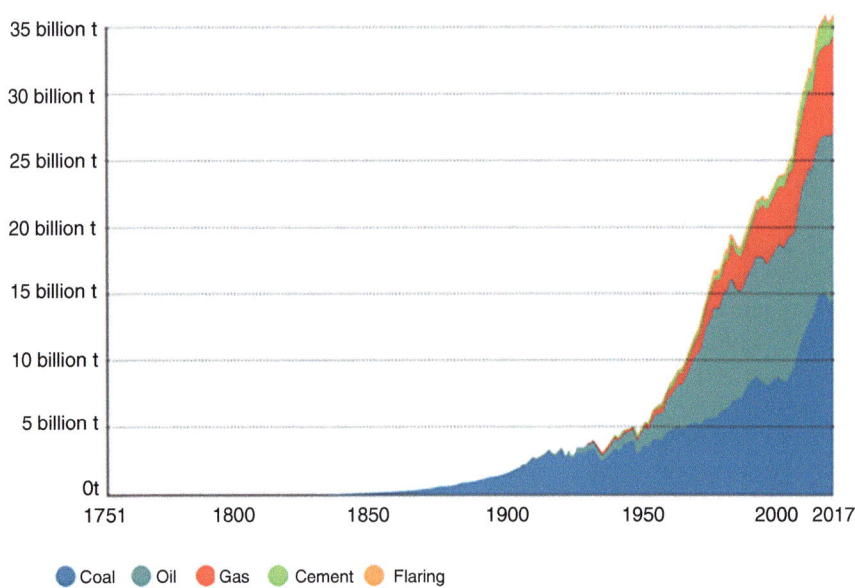

Fig. 8.1 CO_2 **emission by fuel type**. Annual carbon dioxide (CO_2) emissions from different fuel types, measured per year. Cement is included due to the significant GHG emission

land have been almost depleted now and oil extraction from the underground is becoming more difficult and more expensive. Unfortunately, coal is still easily accessible and still exists in large deposits around the planet. The dominant coal-extracting countries are China (46%), followed by the USA (10%), Australia, India, Indonesia, and Russia, each around 7% of the nearly 8000 Mt coal, that is, extracted every year. In Europe, only very little coal is extracted. Germany announced a stop of coal extraction and decided to shut down coalmines, this is now reversing due to the war in Ukraine. Poland accounts for just over 1% of global coal mining. This is a significant part of Poland's economy, but a very small part of the EU economy. The EU should jointly put an end to the extraction of coal as soon as a plan is ready to substitute the gas supply from Russia. Not only for the benefit of Poland but in the perspective of the ambitious climate targets on climate neutrality in 2050. It is worth noting that the outsourcing of European and North American industrial production to Asian, especially China was a relocation of coal as an energy source and a postponement of the transition to more climate friendly and secure solutions.

The reason why it is important to look at the different types of fossil fuels because there is a significant difference in GHG emission per unit energy (kWh) from the different sources of energy as shown in Table 8.1. Here CO_2 emissions from the energy sources based on lifecycle assessment (LCA) are compared.

This shows that twice as much CO_2 is emitted when burning coal than natural gas to get the same energy effect. Which makes it even more important to find alternatives to coal and heavy oil rapidly. Burning coal and heavy oil creates heavy particle pollution compared to natural gas, also a reason for a rapid transition away from the most GHG intensive fossil fuels.

Table 8.1 GHG emissions by electricity source based on Lifecycle Analyses (LCA) to identify the most climate friendly. *Compared by Benjamin & Sovacol*

Technology	LCA estimate (g CO_2/kWh)
Wind, offshore 2.5 MW	9
Hydroelectric, reservoir	10
Wind, land 1.5 MW	10
Biogas—anaerobic	11
Hydroelectric, flows	13
Solar panels, water	13
Biomass	14–41
Solar panels, PV	32
Geothermic 80 MW	38
Nuclear Power	66
Natural Gas	443
Full Cell (hydrogen from gas)	664
Diesel	778
Crude oil	778
Coal without scrubbing	960
Coal with scrubbing	1050

Nuclear Power

An alternative to fossil fuels has for many years been nuclear power but the Uranium-based nuclear power creates the same political insecurity and supply problems as fossil fuel. Uranium exists only in very few places on Earth and often not in democratic countries. Future extraction of Uranium in Greenland is still very uncertain, and it seems that the people of Greenland are not in favor of extraction here.

Nuclear power plants have become safer over time, but there is still a risk of extensive accidents with large damage to people and environment, as seen, for example, after the tsunami in Japan in 2011. When it is possible to base nuclear power on more harmless and readily available raw materials, such as Thorium, this technology has great potential.

Nuclear power plants have become "weapons" in wars which is clearly seen in the Russian invasion of Ukraine. Like renewable energy, nuclear power is electricity producing and smaller district plants based on other elements may be more efficient to balance the other renewable sources of energy than the nuclear plants are.

Biofuels

Table 8.1 also shows that the renewable energy sources can reduce CO_2 emissions faster and more efficiently than burning biomass. However, it requires batteries and other storage of electricity to create a stable supply of renewable energy. Today, biofuels are seen as climate friendly by the EU and a solution that may replace fossil fuels. Biofuels have a climate footprint that is higher than many of the renewable sources and carry other environmental impacts on biodiversity and wild nature that also must be considered. Biomass should not be used for energy production and not for high-temperature combustion, as it requires cultivated land and loss of biodiversity.

The demand for biomass for energy has increased since the energy crisis in 2022 resulting in increasing prices. Biomass should rather be used to produce high-value products, such as food, fibers, lipids, starch, and proteins instead of energy production. Biomass must replace some of the non-sustainable materials in products of the future, such as single-use plastic. It may make sense for some organic residues to be used for biogasification if they cannot be reapplied in food or organic material streams. Then nutrients will be preserved, energy harvested more efficiently, and nutrients can be returned to fertilize agricultural soils.

Bio-gasification is comparable to efficient composting, as we know it at home in the compost heap. In industrial plants where heat and gas are utilized for energy supply, and the residue contains plant nutrients and can be used as fertilizer. This type of fertilizer can be used in organic farming today.

Batteries

Batteries are important in the green transition and renewable energy requires storage of power, including storage in batteries. Lithium batteries have changed the landscape for storing power and have kicked off the development of new types of batteries based on other metals and on solid-state materials.

Batteries are often criticized for their environmentally harmful impacts when extracting Lithium, Manganese, other rare soil, and metals. The environmental footprint of Li-ion batteries is considerably smaller than the general understanding, precisely because they have a large reuse and recycling potential. The batteries can be reused in various future installations, and the metals can be recycled and reloaded. Manufacturers and customers of these batteries have focused on sustainable and responsible supply chains going upstream to the extraction of lithium, manganese, and the other substances used in the batteries. The batteries will be used in many places, such as cars, houses, and at production sites. The biggest battery environmental issue now is the waste of smart devices with Li-batteries and the lack of return of worn-out smart devices and computers from private households.

Research is being conducted on using other metals to make batteries even more efficient, cheap, and safe. The future offers new types of batteries for storing power on a large scale, and here the new "solid state" battery is expected to make batteries more efficient and the environmental impact much less (Råstoffer, 2017).

P2X—typically Power-2-Hydrogen—is an alternative to batteries where electricity is transformed to storage with hydrogen. It may also be turned into other forms as biofuels or heat for storage. The technology is very promising and will become an important storage of renewable energy produced in the future. Large projects are conducted all over the world to develop this technology in many different forms and the solutions for storing renewable energy will be many to meet the consumption demands.

Energy Prices in the Transition to Renewable Energy

The cost of the transition to renewable energy is often discussed and now the prices for renewable energy (RE) are competitive to all fossil fuels. Prices of renewable energy (RE) should drive the transition as this now will solve the recent increase in electricity prices. EU has published an estimate of prices for RE technologies compared to fossil fuels technologies now and in 2030. This is compared with the electricity prices in the EU. Prices on energy including electricity went up tremendously in 2022 due to the war in Ukraine but have stabilized again in 2023 at pre-war prices, noting that the electricity prices were on the rise already in 2021, before the war. The consumer prices including taxes in March 2023 compared to March 2021 have increased threefold.

The way out of this economic and political instability created by fossil energy supply is a transition to renewable energy. Price predictions are important here to create trust in the economic consequences of this transition not to jeopardize consumers or businesses. The price development seen the recent years on electricity with low prices on power in situations with high RE input confirms the conclusions from the analysis presented in Fig. 8.2. A transition to RE will meet the existing price levels or even lower the future prices of electricity.

As seen in Fig. 8.2 renewable energy prices are lower than any fossil fuel except coal, that in the meantime has increased. For the existing capacity of nuclear power prices are still attractive compared to RE, but investments in new power plants to extend capacity will increase the price on future electricity here.

Figure 8.2 contains two graphs—the graph on the left side shows historical and expected electricity prices in 2030 based on the energy mix needed to achieve the EU's climate objectives. It shows that prices are expected to rise only slightly compared to historical average prices before 2018, despite the investments needed in

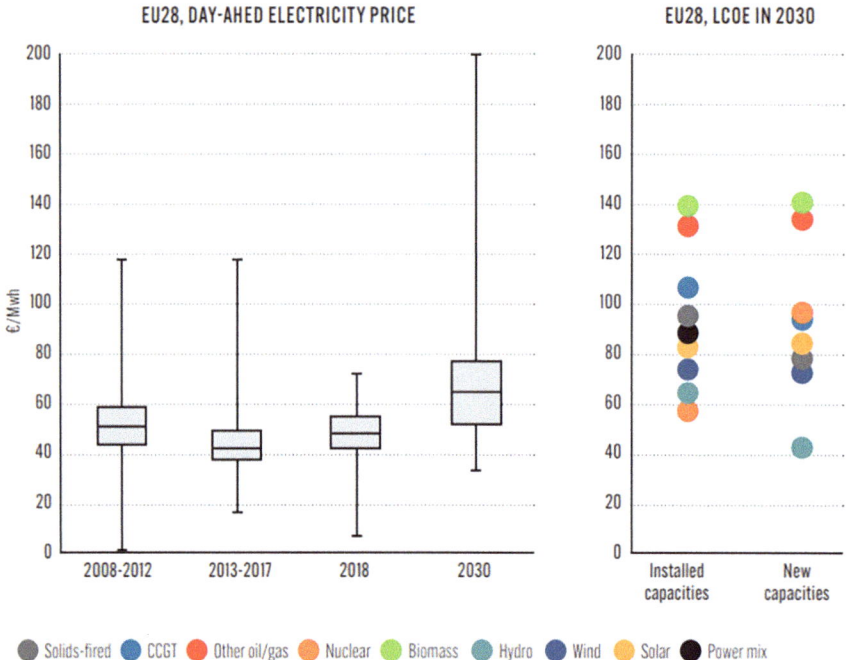

Note 1: the box plots show the minimum observed in a given period (lower whisker), the first quartile (lower bar), the median (black line), the third quartile (upper bar) and the maximum (upper whisker).
Note 2: for visualisation purposes the left graph has been capped to 200 €/MWh.13
Note 3: costs for storage and additional interconnections are not accounted for in this Figure.
Note 4: historical prices are in current euros, values for 2030 are in 2013 euros. Prices and costs are averaged over the EU28.

Fig. 8.2 Energy Prices. Levelized costs of energy (LCOE) are the estimated costs of utilizing various types of energy, including both investment and operating costs over the lifetime of the installation

renewable energy installations and infrastructure. The EU27 average electricity price in March 2023 is around 100€cent/kwh compared to the estimated price on renewable energy in 2030 of around 65€cent/kwh. This leaves space for the transition and even the potential for a decrease in energy prices from renewable energy.

In most EU countries, energy taxes and VAT are quite high, which means that the increase in the price of electricity itself will have a small impact because the price for harvesting energy will represent a smaller share of the total price including taxes.

The second graph in Fig. 8.2 (right side) shows the production cost of energy from the different sources. The costs of biomass and oil/gas are the highest, while those of hydro, wind, and solar energies are below the current energy mix in the EU. RE is driving down the current prices of electricity in these estimated scenarios and seems also to do so in reality. Over time costs will be further reduced. For the new capacity to be installed, the distributed costs of renewable energy are all cheaper than the alternatives, while biomass is most expensive.

The high price on biomass is probably due to an increase in the prices of biomass and waste, due to the switch to biofuels and demand from incinerators. Lower volumes of both waste and biomass will be available for energy production because these resources must be used in the Circular Economy. Introducing the circular bioeconomy means a better and more efficient utilization of the organic (biological) materials that come from forestry or residues from food production. Prices for biomass and waste are already rising, so the assumptions made here should hold true.

Hydroelectric power is cheap, and prices are estimated to decrease further because it is a well-known and efficient technology. Hydropower is today the most widely used renewable energy technology but is not necessarily sustainable on all parameters. The construction of hydroelectric power stations and the diversion of water flows often lead to major changes in nature and flooding of large areas of wild nature. The changes to streams and lakes that hydropower causes are one of the main reasons why one-third of freshwater fish species are endangered (EEA, 2020). The shortfalls of hydropower also became clear with the break of the Hydropower station in Ukraine—again energy supply used as a weapon in wars.

Renewable Energy

Renewable Energy is a supply of inexhaustible sources owned by everyone and is therefore easier to harvest and manage, both in terms of supply and financially. Renewable Energy is probably the strongest democratizing tool humans have had for centuries. In the attempt to spread democracy globally since World War 2, it is strange that equal and direct access to energy has not been the most important tool. Renewable energy is a technology that can operate both decentral and central. Decentralized with modest infrastructure and ownership right down to household level. This applies to both solar panel (EV) and small wind turbines. These small-scale installations can operate effectively around the clock if supplemented with small, affordable battery units. This will empower millions of people in regions that currently have no or little energy supply.

With renewable energy industrial production can go off-grid and obtain cheap and stable energy supply locally in areas where companies do not normally establish themselves due to unstable supply. On a large scale and at a national level, it is of course necessary to control and balance electricity from wind turbines, EV solar panels, hydropower, and storage with the varying consumption over the day, and this requires an upgrade of the electrical infrastructure and storage capacity across countries and regions. On both large and small scales, renewable energy supply is efficient and with the potential of local function and control, based on an inexhaustible source of energy. In comparison the supply of fossil fuels, that is, a centrally controlled energy supply based on a finite source and thereby subject to volatile prices and uncertain supply.

Renewable Energy delivers power when the wind blows or the sun shines, whereas fossil energy production is on-demand supply of energy. With nuclear power being not so on demandable though and then the phasing out of fossil fuel and combustion calls for storage of energy. Innovation and investments in the storage of energy in new ways are required to run (Vedde, 2019) cities and societies at the same level as today and the way we want to develop our cities. The mobilizing of capital for investments in both RE and infrastructure has started and will facilitate the Green Transition. The technologies for renewable energy supply and storage will only become more competitive as technology develops. Unfortunately, electricity prices are still affected by public bills and taxes, and therefore it requires long-term political will and legislation to speed up the need transition to a green and clean economy independent of fossil energy.

> We need to upgrade the electrical infrastructure across countries and regions to fully balance the energy we harvest from windmills, EV solar panels, and hydropower on land and onshore with the consumption on a larger scale.

It is estimated that the energy consumption of Europe can be covered by approximately 1% of the area covered by solar panels (EV), although some researchers believe that the European need can be met by solar panels covering as little as 0.5% of the area. It corresponds to approximately half of Denmark, or less than Jutland's area, to cover the entire EU's needs. This should be a manageable task. The area of 0.5–1% is a considerably smaller area than that of roads, road barriers, parking lots, car parks, and ugly stores in the EU. There is no need to use precious natural or agricultural land for solar panels when we are in such a need for wild nature and biodiversity. Integrated solutions in roofs, facades, existing roads, parking spaces, and car parks are probably a bit more expensive than field solutions, but it should be possible to take this into account with the low COP of solar panels. At the same time, a large share of the power consumption is already covered by wind turbines and hydropower, which could also have integrated solar panels. So, we must prioritize and regenerate scarce nature and place renewable energy technologies as an integrated part of people's lives and the land we already use so abundantly, because renewable energy needs very little space.

In Denmark, a large part of the heat in cities is supplied by district heating (waterborne), and major changes are needed to switch to electricity-based energy. The transition to a 100% renewable energy supply requires the expansion of electrical infrastructure, the creation of new storage, and interaction between different technologies. Energinet Denmark is one of the world's most intelligent infrastructures for electrical supply, and is well-prepared for the future renewable energy supply and the management hereof.

EU are unifying forces and trying to combine legislation, taxation, infrastructure, and private investments with competencies broadly to get the transition seriously accelerated. Still, lack of knowledge and a lot of prejudice are brakes for the green transition, and often people believe that the profitability of the renewable energy supply is too good to be true and become skeptical. It is also frightening to have to accept such large, sudden changes, especially if we should have known this for decades. However, many regions already have a significant supply of power from wind turbines and solar panels and plans for more investments. We are on the way to the conversion of our energy supply, as the easiest and accessible part of the green transition.

Smart Cities and Digitalization as a Tool

Digitalization, as the Internet, new ICT (Information and Communication Technologies), new sensor techniques, new ways of digital labelling, and increasing automatization through robot technology and Artificial Intelligence (AI) are already an important part of transforming energy supply, and an important part of the green transition. These technologies provide access to new types of data on people's behavior and products on a very detailed level.

> Smart Cities are cities driven and controlled by the new digital and sensory technologies. This is how cities are made sustainable because consumption is managed and minimized.

The development of Smart Cities is still immature, and the concept is tested and developed in many places around the world. The transition to Smart Cities will be smooth, and many cities are currently proclaiming themselves as Smart Cities. New ways of recording and tracking people's behavior will be an integral part of the way we adapt cities and industrial areas to the changing climates with increasing water levels, increasing showers, heavy rains, and drought. Cities need much more flexibility and climate adaptation already when they are built and when new urban areas are designed. That is why Siemens and IBM have made Smart Cities their future business strategy.

Digitalization also links to the Internet of Things (IoT) since there will be sensors in all the devices we use and are surrounded by. The devices will report on our behavior as seen with smartphones now. Additionally, all products will contain a tag that will be able to deliver information on contents, origin, sustainability, and coordinates. This will become an important part of the Circular Economy providing real-time access to used materials and products needed at the right spot as well as products ready for refurbishment and redesign to abolish waste and to remanufacture from the existing.

Methods for companies to transform to renewable energy supply and energy optimization are presented in part III of this book, together with sharing of experiences and transformation to sustainable, circular construction is reviewed.

Energy Consumption in the Full Value Chain (Scope 1 + 2 + 3)

Europe has experienced growth and development of welfare since WW2. Europe has ambitious goals of becoming climate neutral in scopes 1 and 2. Though, it is interesting to compare European GHG emissions with global GHG emissions to illustrate the great difference between scope 1 + 2 and scope 3 and in the light of where products are produced and where they are consumed.

Figure 8.3 illustrates the global GHG emissions by sector (Working Group III, IPCC, Edenhofer et al., 2014). It is a different and more simple illustration than Fig. 2.6 in part I, showing the same. The global energy consumption is more clearly divided into sectors here and illustrates that a large part of the energy consumption goes to industrial production which is the energy consumption that is related to the

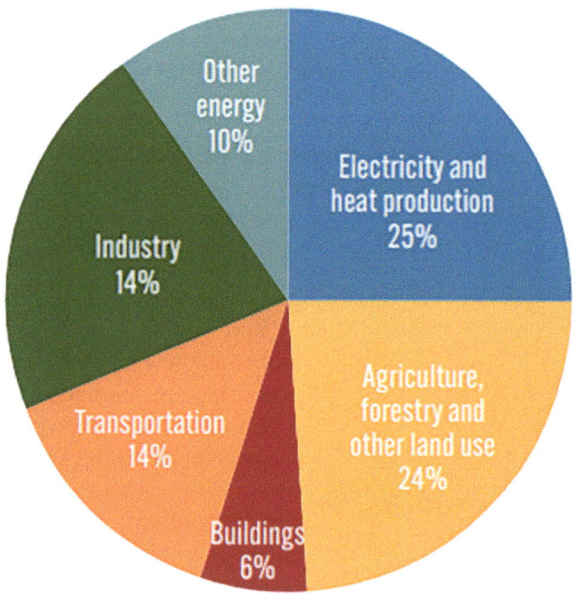

Fig. 8.3 Global GHG-emission by sector. The emissions from sectors in EU from all sources of GHG—energy consumption, production, etc

indirect energy through our consumption of products and is the consumption that relates to scope 3.

Industrial Production and Global Value Chains

Figure 8.4 shows the development of global CO_2 emissions since 1970. China's emissions have increased since 1990 in line with the outsourcing of production from Europe and North America. This graph clearly shows why it is a challenge in reaching global agreements on CO_2 reduction. The historical emission of CO_2 (not total GHG) shows the development that has happened in China over the years. Since 2000, Chinese emissions have increased dramatically due to outsourcing of production from Europe and North America to China. This graph clearly shows the challenges of entering into global agreements on CO_2 reduction. The poorest countries and China have had the approach that consuming countries carry the greatest responsibility for reducing CO_2, while the rich countries do not want to face the loss of growth that this would mean to them, especially at the time when renewable energy was more expensive than today.

Figure 8.4 illustrates GHG emissions directly linked to the geography (scope 1 + 2) of the region and the need for an energy transition in the EU because the main

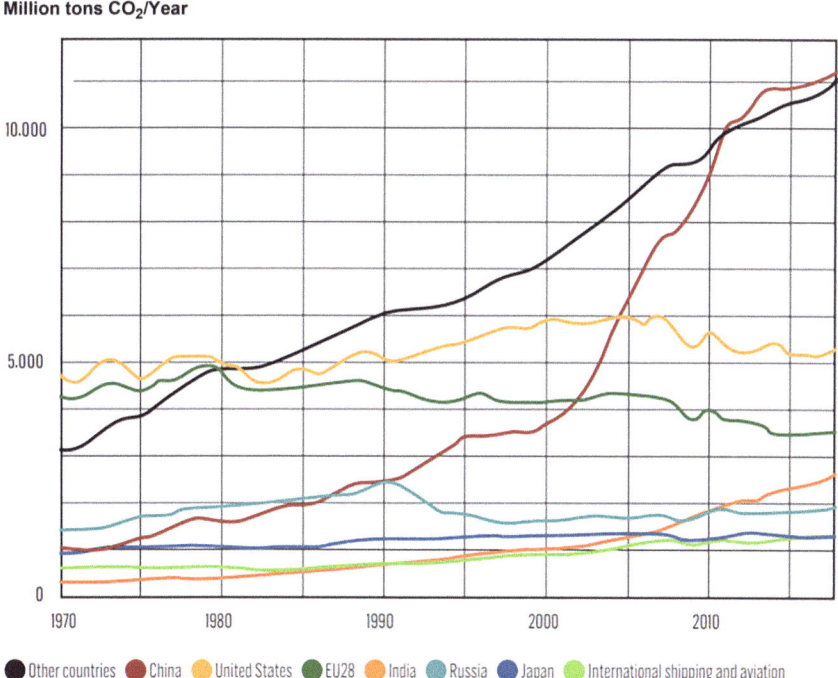

Fig. 8.4 World fossile Carbon Emissions - 1970 to 2018. Development of global CO_2 emissions by large countries and regions to illustrate the shift related to production from the old, industrialized countries to China

challenge is a shift toward renewable energy. The GHG emissions caused by our consumption of imported goods from scope 3 are huge and are not included in these numbers. It accounts for up to 50% of total emissions from the EU (EC, Circular Economy, definition, importance and benefits, 2015). This is why the transition to a Circular Economy is as important as the energy transition and EU Circular Economy Action Plan, described in Chap. 6, also requires a focus on the GHG emissions caused by consumption of imported goods.

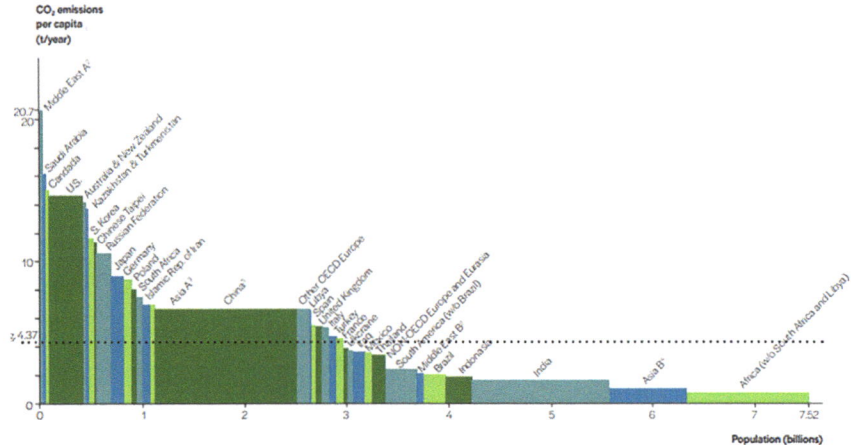

Fig. 8.5 Worldwide CO_2 emission by region per capita (2017). Inventory on CO_2 emissions per capita illustrates that consumption and manufacturing effects country emissions and the need for accounting for scope 3 emissions per capita to drive the urgent change needed from overconsumption. The dotted line is the recommended CO_2 emissions per capita, and a large part of the global population lives below the recommended climate footprint. Based on IEA (2019) world CO_2 emissions from fuel combustion. www.iea.org/statistics. All rights reserved by ADAL Capital GmbH and Tom Schultz. This map is without prejudice to status of or sovereignty over any territory, to delimitation of international frontiers and boundaries and to the name of the territory, city, or area. This work is licensed under Creative Commons Attribution-share Alike 3.0 Unported License. Notes: Energy-related CO_2 emissions only, no other GHG. 1: Middle East A (Bahrain, Oman, Kuwait, Qatar, UAE. 2: Middle East B (Israel, Jordan, Lebanon, Syrian Arab Republic, Yemen). 3: Asia A (Brunei Darussalam, Malaysia, Mongolia, Singapore). 4: Asia B (Asia w/o Asia A, China, India, Thailand, Chinese Taipei, Indonesia, Korea, Japan). 5: China. Version: Dec 17, 2019

If emissions instead are calculated per capita (Fig. 8.5), the large differences between countries become clear, and here are also some of the explanations for these differences. Today, China has also set climate goals, and the USA is again committed to the Paris Agreement. With declarations and goals from these two countries and their collaboration on Climate since April 2021, climate neutrality has fortunately become a global agenda. Mostly because they experience dramatic consequences of climate change these years.

Figure 8.5 shows that countries extracting fossil fuels have high GHG emissions per capita. These countries often also have a high GDP per capita. There is a connection between access to fossil fuels and prosperity over the last 70 years. However, this wealth is not always distributed equally among the people in the countries. Figure 8.6 shows this link as a BNP map.

Russia has not been able to create wealth for its population from its large gas and oil reserves, nor from all the other raw materials that Russia has underground. Australia, on the other hand, has been able to utilize their access to fossil fuels and transfer it to wealth among people. Other large oil nations that are worth noticing are Nigeria and Venezuela, who have also failed to exploit their reserves for the benefit of the general population. Looking at the GDP map, one understands the global political instability that we are experiencing now. Given the significant correlation between energy supply and prosperity, it is time to transform to far more accessible and democratic forms of energy. Many of the raw materials we depend upon are found on continents like Africa, northern Asia, and South America, and some in China. Whereas Europe does not hold many raw materials for the future unless harvested from waste.

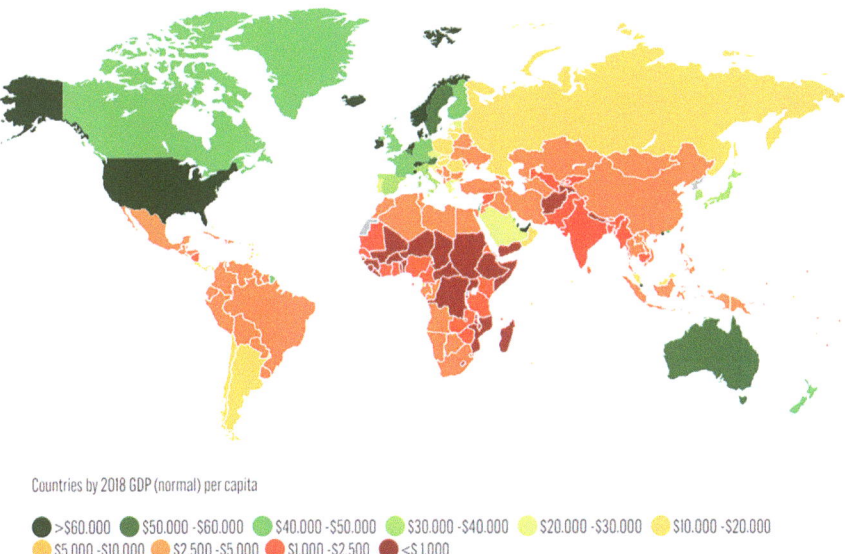

Fig. 8.6 GDP Map per Capita. The spread of GDP per capita is important to investigate when identifying the consumption patterns and economics on actions for driving climate neutrality

All the elements on energy transition are linked to the transition to a Circular Economy that will minimize the GHG emissions in scope 3 of high-consuming countries, at least upstream from the supply of our goods consumed.

Climate Nexus for renewable and efficient energy usage

Apart from the obvious climate effects of zero carbon emissions a transition to an efficient and renewable energy also solves other challenges, as:

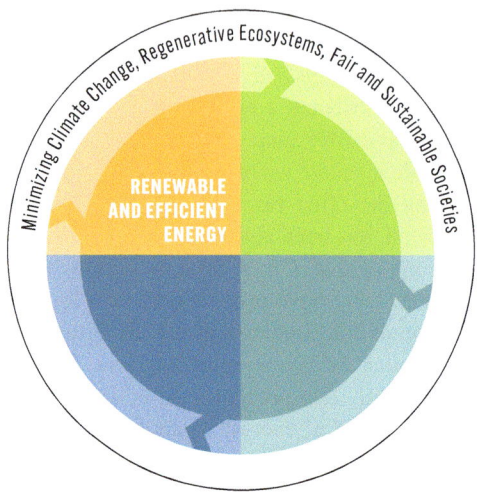

Democratic and free access to energy for all resulting in political stability, spread of democracy, and minimizing poverty.

Predictable prices on energy supply as an important resource for consumers and business.

Eliminating pollution from combustion engines That causes sever health care problems, from particles and noise.

References

EC. (2015). *Circular Economy, definition, importance and benefits.* https://www.europarl.europa.eu/news/en/headlines/economy/20151202STO05603/circular-economy_definition_importance-and-benefits:EU Commission

EEA. (2020). *State of Nature.* European Environment Agency.

Råstoffer, V. F. (2017). *Råstoffer til batterier. FAKTABBLAD om Råstoffer.* Cph: GEUS.

Vedde, J. (2019). Spørg fagfolket: Hvor mange solceller kan dække Danmarks energibehov. *Ingeniøren.*

Working Group III, IPCC, Edenhofer et al. (2014). *Climate Change. Mitigation of Change. Contribution to the Fifth Assessment Report of the Intergovernmental Panel on Climate Change.* Cambridge University Press for UN.

Chapter 9
Transition to a Circular Economy

This chapter on the Circular Economy is a cornerstone in this book and important in understanding the great transition to a Circular Economy. The Circular Economy completely changes the market conditions, the consumption, and the businesses. This chapter is extensive and explanatory to give a guide and a picture on how the Circular Economy looks in the future. This chapter describes the principles of the Circular Economy in details, how businesses are the transformers to a Circular Economy, how new material loops are created, and how all stakeholders are important in creating the great transition to a green and circular economy. Businesses and consumers go hand in hand to create the sustainable consumption that many in the old, industrialized world are aware of the need for.

As mentioned earlier, the linear economy, and the long, global value chains have created large challenges other than climate impacts. These challenges can be summed as:

- Resource scarcity is due to continuous mining of virgin resources without taking care of the basic economy through efficient use of material resources.
- Waste causes environmental disasters on land and in oceans.
- Linear products with very poor quality are designed for waste.
- Lack of control of chemical pollutants due to immediate waste and lack of transparency.
- Unstable and insecure value chains supporting a global economy resulting in a very uneven distribution of wealth and vulnerable economies.
- Lack of transparency and traceability to ensure sustainable consumption.

All these challenges are the reason for the urgent need for a Circular Economy and why legislators in EU have this as a high priority, as described in Chap. 6.

The Principles of a Circular Economy

The main goal of Circular Economy is to maintain as much value as possible in the products and the materials. This is to utilize the materials as long as possible regionally since access to raw materials and components is challenged now and in the future. Circular Economy creates new market conditions, new market potentials, and the need for a change of strategic focus because companies are forced to adapt to new legislation and new consumer demand. The major challenge for companies today is to stay relevant in the Green and Circular Economy. The essence here is to harvest as much value as possible of what is today categorized as waste or used for energy production.

> **The principles of the CIRCULAR ECONOMY are:**
> - Long life and durability are the best and most efficient ways of maintaining value and reducing the consumption of resources.
> - Design for disassembly of the materials with high-quality mono-materials. Products must be designed and produced so that the different types of materials can be separated after end-use of the product, thereby promoting recycling.
> - Design for repair and maintenance to ensure reuse of the products without a need for remaking them, for them to live as long as possible as the original products. Then products live for as long as possible as originally made for (reuse).
> - Disassembly of products into components and mono-materials when reuse is no longer possible, for usage in other products. Then materials can be recycled in cascades making them available for new products. Recycling of components must also be made available for new products.
> - Full separation and sorting of materials for them to be recycled at material level.
> - Ensuring that materials contain no harmful chemical components.
> - Life Cycle Analyses (LCA) on product category level to enable the sustainable environmental choices.
> - Product and material tagging (digital) to facilitate traceability throughout recycling, and to include material history to ensure best possible recycling at the various material banks.
> - New clean material loops and material banks to promote reuse and recycling.

All these principles lead to new types of eco-labelling of products and materials with the Digital Product Passes (DPP) and digital material passes to ensure traceability and transparency throughout the full lifecycle of the product and materials.

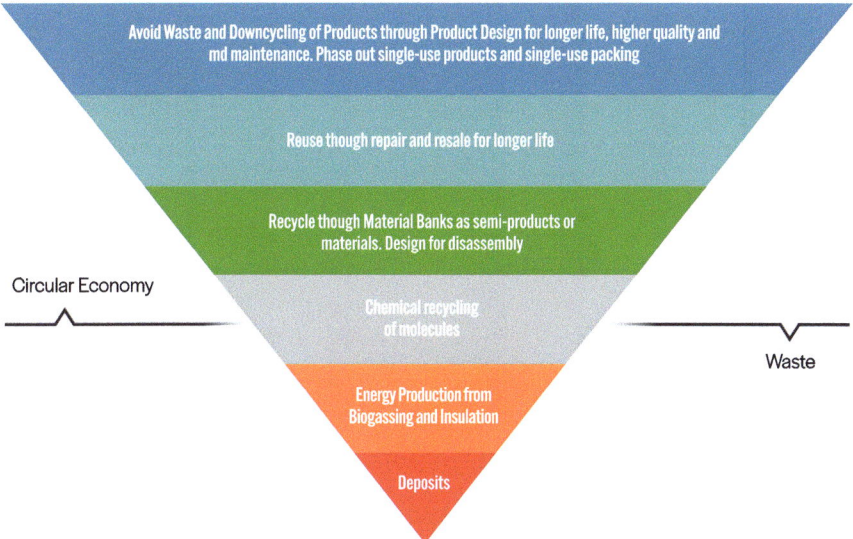

Fig. 9.1 Resource hierarchy

The lifecycle of the materials will be included in the eco-labelling to ensure the best collection, sorting, and quality at the material banks.

The principles of product design are illustrated in the resource hierarchy in Fig. 9.1. The size of each field indicates the amount of value that can be conserved.

The goal is to keep the products as high as possible in this hierarchy to conserve as much value as possible. The waste economy will be prohibited in the EU and legislation is built to design products, services, business models, and infrastructure to start from the top of this inverse pyramid—first the design of good products, then reusing as long as possible before recycling. First mechanical recycling before chemical recycling. If this hierarchy is adhered to a genuine Circular Economy restoring material values as long as possible will become a reality.

Linear Economy

The linear economy is an exception from how the global economy and resource management have worked pre-industrialization. The linear economy is a result of the industrialization and the open and unregulated access to raw materials. It has evolved after World War 2, and it has accelerated in the Western world since the 1980s, particularly since the millennium.

Unlike in the Circular Economy, the linear economy forces a continuous increase in consumption of raw materials to constantly generate the growth on which our societies depend. The linear economy does not take care of resources and leads not

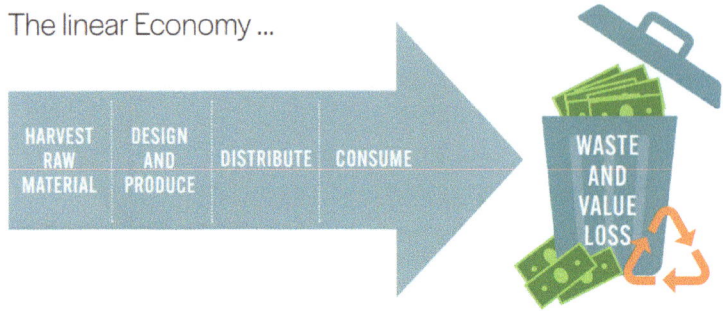

Fig. 9.2 Linear economy

only to overconsumption, but also to losses of enormous values tied up in products when they become waste. The linear economy is simply illustrated in Fig. 9.2.

The whole idea of the Circular Economy is to get away from the production of waste and the loss of values that the linear economy embeds. The Circular Economy is reuse and recycling of resources instead of a constant dependence on virgin raw materials.

Upcycling Is Not Always Circular

Now, we are seeing many new types of upcycling of waste to products. Upcycling belongs in the Linear Economy and is what the upcycling industry has lived from for many years. Upcycling is not bad in a situation with a lot of waste available, although often upcycling is not circular. The recycling industry also needs to be innovative and play a new role in the Circular Economy, then new business opportunities are available here. If we extend the linear economy with more upcycling, we will not reap the great economic and environmental potential of the Circular Economy.

It is important that the circular products have an afterlife when they have finished being used and reused as designed for. This also goes for new, reused, and upcycled products. It requires that the upcycled products live up to the principle of the Circular Economy (highlighted):

- Design for disassembly.
- Products must contain no harmful chemicals.
- Maximum potential of reuse and recycling through maintenance and repair and recycling loops.

A lot of upcycling of waste for building materials and clothes-like products is occurring at the moment. A product like MDF plates is one example of an upcycled product with limited durability, no repair potential, and no design for future recycling. At the utmost to be incinerated for energy or put on landfill. When shoes and clothes are made from ocean plastic, that may be a way of getting plastic out of the oceans. It is not circular, unless the shoes, clothes, and materials are designed for a valuable afterlife. The shoes should be made for disassembly into nye monomaterials and the plastic should hold a quality that is recyclable. A circular product is not only made from recycled materials it is also made recyclable in the future and with a reuse potential. Not all upcycled products comply with this.

Thus, we must not just extend the linear economy and develop more recycling from waste or new upcycling technologies that require a continuous generation of waste. Then the potentials of a Circular Economy disappear.

One example that illustrates the problems created by the linear economy is plastic in the oceans and microplastic spreading in the environment. Microplastic originates from the plastic fibers used in textiles and weathering of plastic from textiles and car tires, but also from other products. The fractions swimming in the oceans are blended and damaged plastic polymers, which cannot be recycled in the recycling systems available for plastic based on monopolymers. The producers must take responsibility for the recycling of these "new" ocean-plastic products after end-use or new problems are created or the falls of the linear economy are just postponed. Technology is available to chemically recycle this mixed and damaged plastic. This will solve some of the challenges, but it is still important to contain the highest possible value, and only then profitable recycling systems are created.

Circular Economy: The Manufacturing Industry

According to Ellen MacArthur Foundation's report: "Completing the picture" (2019), the climate impact from our consumption of products covers 45% of the global GHG emissions as seen here. These aspects are also highlighted in other publications.

The building industry shows a very good example of the Climate Nexus to our consumption since various analyses state that up to 70% of the GHG emissions from this industry come from the production of materials and construction, and only 30% from the energy consumption in the buildings. The building industry and the buildings cover approximately 50% of national GHG emissions in many countries, including Denmark.

Putting energy savings in a circular context means that we need to be careful when setting energy-saving targets and introduce a long-term life cycle approach when assessing the GHG emissions of buildings. EU has focused on energy savings for some years. Now large-scale energy-saving schemes are initiating a huge wave of replacement of windows, light fittings, roofs, etc. In most cases, only the energy saved from the consumed energy in the use phase is accounted for. Not the energy (GHG) wasted due to replacement of products that also require energy to produce.

The business model of take-make-waste consumer goods has also reached the building industry, and new products are often of poorer quality and with no possibility of maintenance and repair, the only future prospect being repeated exchange of products of low quality. This goes for windows that today are often made of plastic or for light fittings where LED has become the ruler of energy savings. LED is a brilliant technology with so many benefits including light quality, but it does not justify installing single-use fittings of the light source (LED), which cannot be replaced, and the whole fitting needs replacement after 3–5 years, when the LED has leaped.

Large providers like IKEA and Phillips are using this business model, and that is reprehensible. Phillips introduced the business model of delivering light instead of fittings (service model), which resulted in the installation of many single-use fittings because the LED is built into the fitting directly instead of being detachable. This has also resulted in a huge waste of old, classical high-quality fittings made of metals that could have been upgraded—because we now see companies with a circular business model retrofitting the existing installations and installing LED, thereby obtaining a GHG saving of an additionally 40% compared to new ones—even new high-quality ones.

By tagging all materials and goods with new digital technologies, transparency, and traceability of materials and content of products will become available to support the Circular Economy and meet some of the challenges we have experienced with chemicals. Digitalization is a very important facilitator of the Circular Economy and keeping track of resources will become very important to ensure efficient access, as described in Chap. 6.

Chemical content in new products tends to be increasingly higher than in old products—e.g., building materials after the 1970s and textiles after 2000 have shown low lifetimes and a mix of materials and chemicals. The use of chemistry to create new and other technical abilities in products has also increased. Far more different chemicals are in products and the so-called cocktail effect from mixing chemicals is increasing.

The reason why the Circular Economy rapidly will minimize GHG emissions is that mining and harvesting virgin resources require much more energy than reuse and recycling of materials. Figure 9.3 shows the energy that can be saved by recycling materials and creating material loops locally, rather than extracting from the virgin mines. Additionally, the recycling industry is easier to electrify than the mining industry and this itself creates energy savings.

Figure 9.3 shows that the immediate energy savings from recycling materials rather than extracting virgin resources are between 30 and 90%. Here the building industry, cement production, steel production, and the extraction of other metals are large sinners. The savings from recycling of cement can be much higher than stated because the recycling technology is premature and used concrete is put as foundation in the construction industry in building roads, etc.

Fig. 9.3 Saving from recycling or other circular concepts (CO2eq). Source 1: Climate Benefits of Material Recycling Inventory of Average Greenhouse Gas Emissions for Denmark, Norway and Sweden. Litterature Study. University of Gävla, 2015. Karl Hillman et al. Source 2: Uddrag fra Livscyklusvurdering for cirkulære løsninger med fokus på klimapåvirkning, Forundersøgelse. Statens Bygge Institut, 2019. Camilla Ernst Andersen, et al.

Material	Saving from recycling or other circular concepts (CO2eq)
Glass[1]	41%
Aluminum, general[1]	96%
Plastic[1]	37-55%
Paper and Cardboard[1]	6-37%
Organic waste, biological composting[1]	21-27%
Organic waste, anaerobe biogas[1]	54-87%
Bricks[2]	61-77%
Concrete elements and pillows[2]	96%
Recycle concrete[2]	0,3%
Steel[2]	78%
Wood, beams, boards and posts[2]	77%
Chipboards[2]	9%
Plaster[2]	10%
Window[2]	77-96%
Roof tiles[2]	98%
Aluminum, buildings[2]	81%
Door[2]	80%
Roofing felt[2]	69%

The Basic Features of the Circular Economy

Circular Economy is simply making the value chain circular and create loops to maintain access to the materials, semi-finished products, and raw materials of our products. The goal is to avoid mining of virgin raw materials and the manufacturing of new products all the time. We already have several products that are ready for the circular loops. Now, legislation is paving the road to a Circular Economy, but business models, infrastructure, and logistics do not support this new economy and we need new ways of manufacturing, consuming, and returning. Especially products of older age, such as cars, furniture, and building materials, can be repaired and maintained, and most of the products can be disassembled for the recycling of materials.

It is especially the increased amount of chemicals and plastic variants that make the products of today not suitable for the Circular Economy. The best example of a circular business model in Denmark is our deposit system on bottles, which for many years only applied to glass bottles, but now also aluminum cans and plastic bottles, and shows very high rates of reuse and recycling close to 100%.

The concept of value chain comes from M. Porter's Value Chain, which is not a value chain but a way of organizing a company in a linear economy, as Porter's value chain only describes the product's journey through the company. Porter's Value Chain is a result of a linear way of thinking and probably contributed to the way we look at companies as silos in a long value chain. Today, companies need to understand the full value chain of their products in a circular way, as illustrated in Fig. 9.4. Learn more about the rethinking of Porters value chain in (Haar, Rethink Economics, 2024a).

> **We can't solve problems by using the same kind of thinking we used when we created them.**
> —**Albert Einstein**

The circular value chain in Fig. 9.4 is illustrated as an infinity sign with two circles because major changes must take place both in the production and service industry (green area) and in the recycling industry (yellow area). Most importantly, is to make maximum use of the products on the right side of the model, and products must live for as long as possible as designed for. This is achieved by reuse, maintenance, repair, and resale.

When products are worn out and cannot be repaired any longer, they move into the recycling circle (left circle), and here materials and sub-components are recycled as furthest possible. Then materials and values are harvested in cascades, and always with the highest possible value conservation. Thereafter, there is often a further recycling of the materials via a chemical process upgrading to raw materials again. Only when the products or materials are completely worn out and have been used in cascades of different uses, also chemical recycling. Then they pass on to energy production. Efficient energy production either in an incinerator or a biogas plant if an organic fraction. For a small fraction, the rest materials are moved to deposits due to the content of hazardous chemicals.

The most important keys to make the two circles and a Circular Economy work are A. NEW MATERIAL BANKS & COMMUNITIES of RAW MATERIALS and C. NEW BUSINESS MODELS, which are seen between the two circles in the infinity sign. Material banks are likely to emerge from the existing recycling industry, and the new business models will emerge in all kinds of industries in both the manufacturing companies and attached service industries. Especially a new material infrastructure and the move of materials from the green to the yellow circle is crucial for the success of the Circular Economy.

The steps in the circular value chain will be explained below to make it clear how Circular Economy and the new business model create value for companies and consumers (Fig. 9.4).

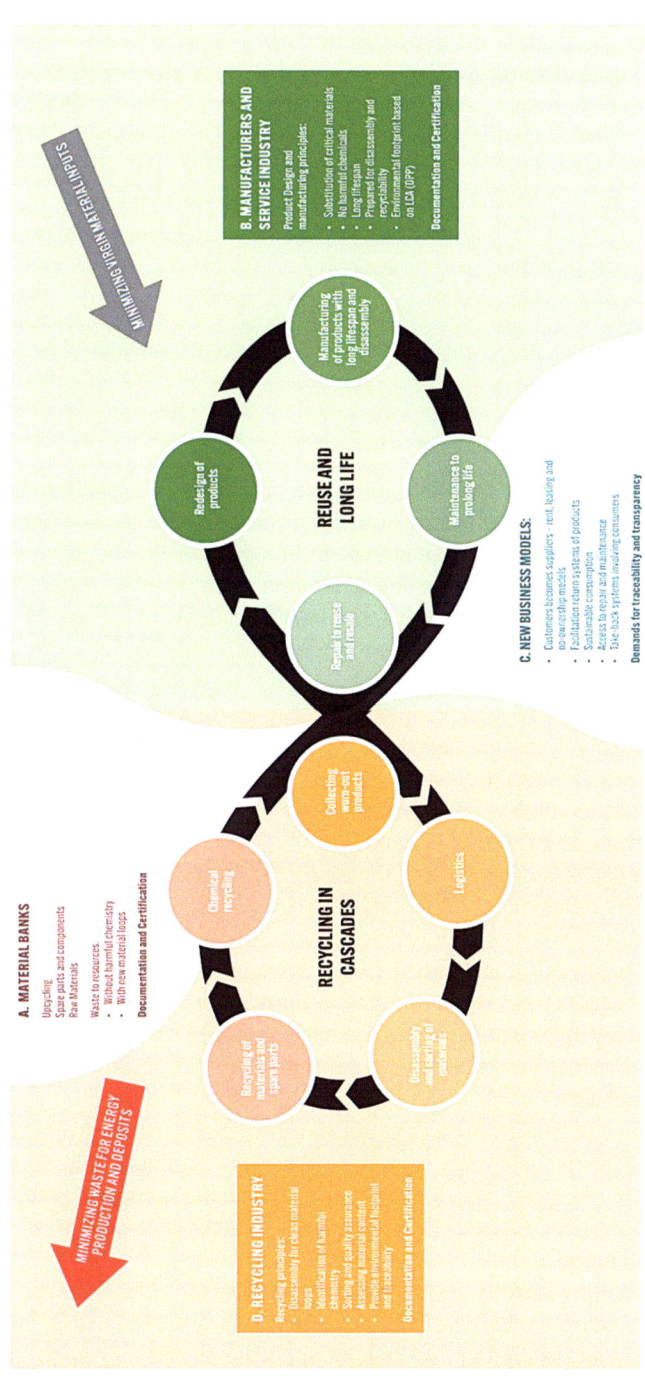

Fig. 9.4 Circular value chain

A. MATERIAL BANKS AND COMMUNITIES OF RECYCLED MATERIALS

In the future, companies will get recycled *"raw materials"* and semi-finished products from **banks** or **communities** of various materials. The material banks may evolve out of the existing recycling industry taking care of some or all the steps in the left circle (yellow), covering; sorting, separation, quality assurance, and description of the materials to make them reusable and recyclable. This requires massive innovation from the traditional recycling industry if they are to take over the supply of recycled materials for the manufacturing industry and contributing to design and development of new products.

> In the future, our raw materials and components come from material banks of recycled materials.

The municipalities are responsible for waste collection and handling in many countries, thereby being the owners of household waste. A need for liberalization of waste handling is necessary for the materials to find their way to the material banks or manufacturers for reuse and recycling. It is important that there are national and international recycling systems and material banks available to all stakeholders in an industry. Liberalization must not result in a few companies or actors completely controlling the various material loops and banks. It is important that there is a standardized and accessible system available for everyone in the industries, including the SMEs. Especially in plastics and textiles, where new infrastructure is to be built, it is important that systems are created accessible to the entire industry ensuring access to materials for all. It is crucial not to create material monopolies and thereby inhibit the transition to a Circular Economy.

The material banks need to account for the footprints of the reused products and recycled materials based on life cycle analyses (LCA). The environmental impacts have to be calculated based on standardized, scientific methods, and validated databases. Existing methods for computing LCA's accounting for the entire lifetime of a product and materials are suitable for this. Introduction of the EU Sustainable Products Initiative (SPI) with Eco-design criteria and the Digital Product Passport (DPP) with the Product Environmental Footprint (PEF) will result in standardized methods and databases.

Implementing a broad understanding of LCA the material banks are able to support designers and manufacturers in making the right material choices, with the smallest possible environmental footprint and the greatest possible potential for reuse and recycling. The Circular Economy includes completely new types of data on materials and products.

Materials will move directly to recycling after use and reuse, and thus the amount of waste will decrease drastically. Only if the products or materials become completely worn out and are used in a cascade of applications will they go to chemical recycling or to energy production in an incinerator. The same applies to organic fractions or biomaterials. The organic residues can be utilized in new contexts, such as fibers, proteins, starch, and so on. Only when there is no more value in the organic material will it be used for energy production in biogas plants?

Challenges may arise since the EU countries do not hold identical recycling systems. For example, some countries use industrial composting at high temperatures where some plastic types can decompose. In other countries, as in the Nordic region, degradation of organic material occurs by biocomposting mixing the organic household waste with large amounts of animal manure, and here plastic cannot decompose. Biocomposting occurs at low temperatures and sets other demands on the input materials. Thus, there is confusion about the biodegradability of plastic across countries.

It has been decided that all EU countries must implement the same sorting standards based on sortering at the source (households, etc.). See more on sorting standards later in this chapter.

For manufacturers reuse and recycling of products and materials also entails challenges. Producers and consumers want to be sure of the quality of materials and products, whether they are reused, recycled, or new. It requires guarantees and insurance at the same level as when buying new. For example, the contractor or craftsman will have a guarantee of the quality of the recycled and renovated building materials, in the same way as when buying new. It must also be possible to return the item in case of mistakes and defects at the same degree as with a new. It is also necessary to know the content and quality of the recycled materials to be able to resell them. Both due to legal requirements, but also because the consumers have a right to know what they are buying. Therefore, the same requirements must be imposed on recycled products as on new ones. This is an important step in securing the transition to a Circular Economy.

The Circular Economy requires clean material flows, and recycling of materials is often challenged by the chemistry added earlier. Recycling often concentrates chemicals and creates problems and limitations for future use of recycled materials. It is a long journey to get clean material streams for the Circular Economy and only with clean materials and security of content in the recycled materials the maximum value can be maintained from recycling without harming the environment or people. See more on chemistry in this chapter.

B. MANUFACTURING AND DESIGN

In future, products must be designed and produced according to the above circular principles to be part of the Circular Economy and the material loops. The new material infrastructure and increased requirements documentation will have a massive effect on companies.

> Design is not the full solution—new material loops pull innovative design solutions.

Designing products for the Circular Economy receives a lot of attention these years and it is a key to better preservation of value when products are designed for long lifespan and disassembly. It is often overlooked that access to recycled materials of sufficient and standardized volumes is at least as important. It is to a larger extend structural changes, new material infrastructure, and new business models at all steps in the value chain that are necessary for the transition to a Circular Economy. A huge number of changes are necessary along the value chains of materials and products and evolve into new circular loops before the Circular Economy is fulfilled. Business models must include take-back systems of products to ensure reuse and recycling, and this again will drive new product designs.

Documentation and Labelling
Documenting the footprints of products will become important, including material footprint, climate footprint, chemical footprint, and footprints on wild nature and biodiversity. All the documentation required in the transition will affect companies enormously and the requirements for these new non-financial business data are the same as known from the financial business data when it comes to traceability of data and transactions, transparency, and auditability.

Unfortunately, today consumers are confused by the existing eco-labels and the increasing number of retailers' private eco-labels. Studies indicate that the increasing number of eco-labels is causing confusion among consumers because it is not clear to them what the scopes of the different eco-labels are. Nevertheless, several new private and public eco-labels will enter the market in the years to come. EU wants trustworthy eco-labelling to create an inner green market and to counter greenwashing. The requirements put on existing eco-labels to map to the EU-PEF will come, and lifecycle analysis will be the only way to document and claim sustainability in the future.

The implementation of PEF is not imminent, but over the next 3–5 years we must expect to find this label on many products, either directly as an EU label or indirectly by lifting the standards for the existing eco-labels. Every product must carry a Digital Product Pass (DPP) based on PEF and other information throughout the value chain. Then products can be benchmarked to make the choice of sustainable and circular products much more confident. Eco-labelling is necessary, but a way through the jungle of eco-labels is needed until EU legislation is fully implemented on SBI, Eco-design criteria, and PEF. See more on EU legislation on products in Chap. 6.

New Refurbished Products
In recent years, a new industry of repair shops and shops that sell refurbished goods has emerged in many countries. Here good products are repaired on behalf of the owner or resold by the shop as refurbished of "reloved" products. Today, lots of products are thrown away, even though many of these have a good lifespan left in them. This applies, for example, to household appliances, electronics, and IT.

Often the physical device itself is not worn out when wasted, often only the electrical/digital systems are unfunctional or outdated. Filters, gaskets, and other minor components that may be worn out can easily be replaced or repaired. Then, most of the products can be used again and again, which goes for cars, home appliances, and electronics, but also building materials, textiles, and plastic devices.

Electronic devices must hold a much longer lifespan at much higher prices than today. Many usable electronics are scrapped today, due to low prices and the lack of financial incentive to repair and maintain. In Denmark, an investigation showed that 80% of all electronic devices that are wasted have the potential for repair and become fully functional again. Wasting good products and especially electronic devices creates major environmental problems, and scarce resources such as metals and rare soil minerals are not recycled. Even if we need these raw materials in the green transition. Especially the transition to renewable energy technologies requires a lot of metals and rare soil metals that we cannot continuously harvest as virgin resources. In the future, more IT manufacturers will offer upgrades of their products and refurbishment, rather than sale of new devices, either as a swap solution, where the product is given back to the manufacturer, or by making a digital technical and software update that reduces resource consumption. Within PCs, tablets, and smartphones, we are already seeing a growing used and resale market, but especially smartphones are continuously upgraded with new software.

Information and Communication Technology (ICT) hardware suppliers are a particular challenge. Apple has been convicted of blocking the updating of old versions and thus, on false pretenses, getting consumers to buy new ones. It is beneath contempt that the potential for recycling and repairing computers today is lowest with the largest brands, such as Apple and Microsoft. Unfortunately, it is a general trend that the big, global brands struggle to maintain their linear business models and their monopolistic markets, while the smaller brands seem more adaptable. Both on the European and North American markets restrictions are coming to counter these monopolies of the large ICT device providers for them to open for repair, maintenance, and refurbishment of product to force longer lifespan.

C. NEW BUSINESS MODELS AND NEW CONSUMPTIONS MODELS

For the Circular Economy to work, new business models must be developed with less focus on linear sales and more focus on preserving the products and materials in loops. Figure 9.5 illustrates some circular business models suitable for certain product types in relation to product lifetime. The illustration is a theoretical presentation and should be seen as an inspiration to rethink the company's business model. Management in every case needs to identify the business model that suits their specific products and that customers can adopt.

When a company transforms into a circular business model, the model must be adapted to the specific products, customers, and markets. As when a company develops its strategy and finds its unique position in the market (Unique Selling Point). This is an ordinary strategy exercise within a circular mindset to meet completely new market conditions both downstream and upstream.

It is a misconception when much literature and many advisors prescribe that circular business models are service models. It's not that simple. In a Circular Economy, a company must identify the markets, the customers, and the spot in the value chain when developing their business models, as usual. The goal with new

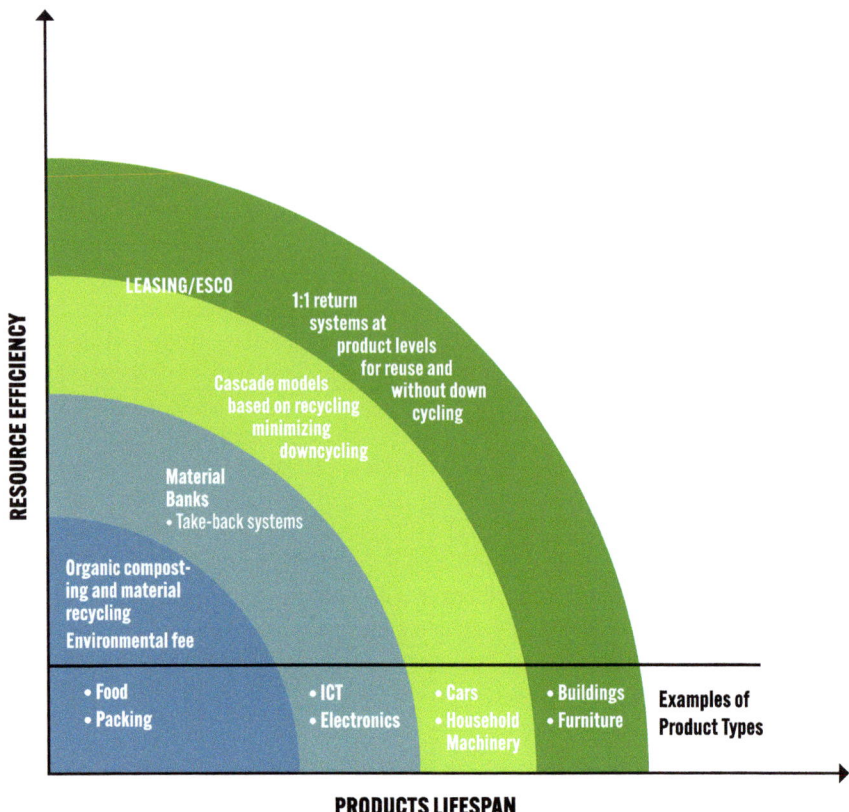

Fig. 9.5 Business Models. Examples of circular business models of different product types. The best-suited business model is dependent on the exact product and the company

circular business models is to ensure future access to products and materials, why companies must include reuse and recycling in their business models and their value chain. This is done with different types of business models depending on the type of product, the company, and the market situation.

The circular business models will bring customers and suppliers closer to each other because everyone in the value chain is part of the new material loops. The new EU legislation also includes extended producer responsibility (EPS) on all product categories. This will affect the circular business models, since manufacturers will become responsible for the extended value chain and for extending product lifespan, ensuring repair and maintenance, and recycling of materials.

The new digital platforms are an important part of companies' circular business models and will often support the green transition. But digitalization may also prolong the linear economy. Digital sales platforms are effective channels with new types of marketing and communication, and often with no or little focus on the environmental impacts of the products and the manufacturing processes. Then, digital platforms contribute to accelerating consumption rather the preserving resources and minimizing environmental impacts. On the other hand, digital platforms can contribute to disruption of the traditional way of retailing and bring the consumer and the manufacturer closer together, for example, as a kind of online "local" farm store. Read more about digital business models and "Sharing economy versus Circular Economy" later in this chapter.

Low Prices and Greenwashing Are Leftovers from the Linear Economy
The hunt for low product prices has become a large problem and has resulted in the linear economy and global value chains. It jeopardizes the labor conditions, environment, and our resources. Stuff has become too cheap and quality to poor. No incentives are left to reuse, repair, and maintain our products when prices are so low. The pure financial incentive of creating low prices enhances the linear economy when it is cheaper to buy new products instead of repairing the existing ones. Therefore, it is important that prices as well as product quality raises as part of the circular business models.

> Higher prices are needed in the Circular Economy, not necessarily increasing costs.

We need to pay the full price for decent and environmentally friendly manufacturing and transport of our products. As well as we need to pay for the impacts of harvesting natural resources and impacts on wild nature, biodiversity, and the climate (externalities). Products with a long lifespan and of good quality often hold lower environmental impacts than cheap junk.

In particular, the public sector and the global corporations have a large share in the pursuit of cheap prices and have thus stretched out the linear economy to an extent, where low prices have become the main problem. Public and private procurements must be carried out on other parameters than just the lowest initial price. Competitive bidding and tendering must be based on quality, sustainability measures throughout the value chain, and the total cost of ownership (TCO) of products throughout their full lifetime.

The EU countries have to a different extend used public procurement as a lever for the green and circular transition. In some countries very little is changing, as in Denmark, despite good intentions, declarations, ambitions, efforts, and working groups. Lack of competence within the circular and green economy as well as the insight into products' total costs of ownership and sustainability is one barrier in countries that are lagging in the Circular Economy. Whereas in other countries, as for example France, Germany, and Sweden they are far ahead in using public procurement as a lever. The same applies to many global companies, where procurement is focused on initial low purchase prices or simple payback time and this will result in short-term decisions on cheap products. In the Green and Circular Economy companies are to take responsibility for scope 3 where it often matters the most.

D. RETURN AND RECYCLING

There is a need for transformation of the recycling industry and public–private waste management in the transition to a Circular Economy to give manufacturers access to recycled and reused materials in new ways. New business models are required in the recycling industry as well as in the manufacturing and service industries. The new business models here will ensure clean material flows and develop future material banks. The recycling industry must be even more specialist in the different types of materials and work tighter with the manufacturing companies in creating material loops and products designed for this.

Taxation is a strong instrument and there are strong indications that the EU are aiming at introducing taxes to facilitate recycling. This will mean that products made purely of virgin materials will be taxed more heavily than recycled products and products not prepared for future recycling. Then there will be a downward taxation toward no taxation of the most circular products. More on this in Chap. 11.

It is crucial that new infrastructure is created for the materials all the way through the two circles in Fig. 9.4. All the way from citizens or companies sorting of waste to collection, handling, post-sorting, and quality assurance of the recycled fractions. Today, the lack of recycling systems and material loops is the biggest obstacle to the transition. This applies to many types of materials, such as organic materials, textiles, and plastic.

Contrary to common beliefs, there are virtually no or very few recycling systems for textiles or plastics in the EU today. Opposite to metals, organic materials, and fiber-based paper and cardboard. Many in the textile industry claim to embed sustainability, and this often is the use of organic, virgin cotton, fewer collections, and higher quality, for the clothes to be used longer. A few enthusiasts are working on the recycling and reuse of textiles (not clothes). Test facilities for recycling of fibers exist, but there are no structural changes to achieve a circular value chain for textiles in Europe, and there is still a long way to go. EU has decided to introduce a separate household collection of textiles as of 2023 and has introduced a strategy for a circular textile industry with strong recommendations but the way to circular loops of textiles is still long.

> Transforming the recycling industry and public waste handling to Circular Economy is quite a task. Thus, it is necessary to give the manufacturing industry new access to reused and recycled resources.

New material banks also mean new logistics. In most countries, much will circumvent the waste collection system as we know it today. The six EU waste directives (2017) prescribe recycling percentages already from 2020 being implemented until 2035—along with the following as seen in Fig. 9.6.

TYPE OF WASTE / RESOURCE	EU REPORTED 2016-17 (OLD STANDARD)*	GOAL 2025	GOAL 2030	GOAL 2035
Total recycling excluding mineral waste (included material recycling, composting and anaerobic digestion)	57%			70%
Recycling of household waste and corresponding waste from companies	46%	55%	60%	65%
Recycling packing waste -With specific recycling goals for packing made from plastic, wooden fibers, metals, aluminum, glass, paper and cardboard	67%	65%	70%	75%
Electronic and electronic devices	41%			
Building waste	70% (2020)			
Maximum deposits of household waste	10% (2030)			
Separate collection of organic waste from households	2023			
Separate collection of textiles from households	2025			
Mandatory Extended Producer Responsibility (EPR)	2025			
All packaging material from plastic must be reusable or recyclable in a cost-efficient manner	2030			
New Standards for monitoring and register recycling rates, not yet fully implemented	2019			

*include export of waste without accounting for final disposal.

Fig. 9.6 EU Target Recycling Rates. The goals from EU waste directives (2017)

Today, products are transported one way, from producer via wholesaler to retailer and consumer. Then waste fractions are transported to different kinds of disposal, depending on the degree of sorting. Resulting in separate logistic systems for the new products and for the used products and waste handled by different transporters. More on the transition of logistics in Chap. 10.

Waste Is History

Over 2151 million tons of waste is produced every year in the EU or 4.808 kg per capita of total waste (2020, EC, Waste Statistics) including all types of waste. Denmark is highest on the list in the EU when each Dane produces an average of 844 kg of household waste every year (2019, Eurostat), while an average EU citizen produces 502 kg of household waste every year (2019, Eurostat).

Figure 9.7 shows how much waste is generated in the EU by country.

Europe is very high in waste generation and the potential for utilizing these resources is huge. 39.2% of waste in EU is reported as recycled and 60% of all household waste ends up in landfills.

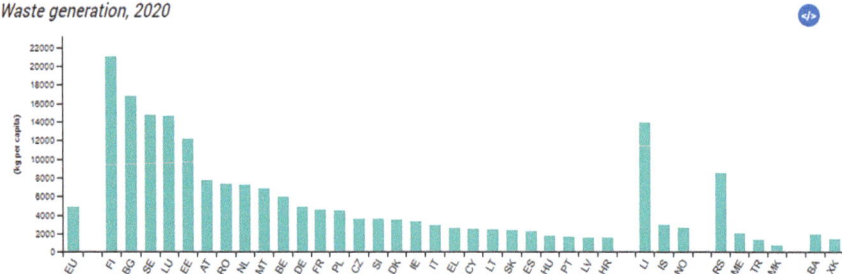

Fig. 9.7 Waste generation in EU by country (2020). KK (Kosovo): This designation is without prejudice to position and status and is in line with UNSCR 1244/1999 and the IC: Opinion on the Kosovo Declaration of Independence. Source: Eurostat (online data code: env_wasgen)

Until 2019, waste was exported out of EU and a significant amount of the waste exported ended up in less developed countries in landfills, in nature, or in the ocean, causing global environmental disasters. The new EU guidelines (2019) for reporting recycling rates of the different resource/waste fractions mean that the exported waste is no longer included in recycling rates unless there is full accountability for the recycling of materials. Official recycling rates will drop dramatically across EU in the future when the new rules on the export of resources are implemented. This will also affect the number above of 39.2% recycled materials. In the longer run actual recycling numbers will increase with the new regulation. Thus, the figures will be in line with the overall estimates of resource circularity of 8%.

Sorting and waste collection in EU have undergone major changes these years. Denmark has adopted a new set of sorting standards for private households, municipalities, and companies. The sorting standards are applicable as of this year (2023). Nordic Council covering all Nordic counties is following the implementation in Denmark and may implement the same standards within a few years and the same goes for EU, since EU already decided on uniform sorting of waste years ago. The waste pictograms and user guidelines has been the foundation for a joint cooperation between several Nordic and Baltic countries, towards a harmonization of the use of pictograms in waste management in the Nordic region. See: www.EUpicto.com as a result of that cooperation. With good and well-defined standards, the road is paved for implementation in EU. Standardized sorting, collection, and handling of all waste throughout the EU will result in significant volumes of uniform materials that can ensure that recycling is profitable and is a key to running the loops in a Circular Economy. As mentioned, a liberalization of household waste is being implemented making waste a resource and thereby worth investing in the infrastructure. All these are main contributors to the transition to a Circular Economy.

Danish Sorting Standards. The total number of standards is 91 and the picture illustrates the fractions of sorting household waste in Danish

The actual solutions and new infrastructure are still to come for some of the materials, as textiles, plastics, and building materials. It is important that profitable solutions can be created to manage these important resources in the future. It is expected that a larger part of the recycling will be done by private companies and corporations between private actors in the same industry.

It is important that the infrastructure of material loops is created and that the resources are accessible by everyone, and that resource or waste monopolies are avoided. Recycling and reuse must not take place in closed entities by a few private actors, as it is evolving for plastics right now. For the existing material streams with high recycling rates, such as glass, fiber-based materials, and metal, monopolizing materials is not a challenge. Recycling new types of resources as plastic, textiles, etc., it may become a challenge that only a few have access to materials in the future. Where infrastructure is lacking the battle for access to recycled resources has started. There must be national and international infrastructure available to all operators and companies, including SMEs. The green transition needs everyone on board, and we should not allow a few large corporations to monopolize our resources and then not create sufficient volume and uniformity to make the systems long-lasting and inclusive.

The largest change is on the "left/yellow side" (distribution and recycling) of the Circular Economy in Fig. 9.4, and that is where the largest innovation is happening now. Once new clean material loops based on recycled materials are available to manufacturers, it will push the changes in design and manufacturing of products needed on the "right/green" side of the model.

A Danish example of a company that creates change and innovation in the recycling industry is Meldgaard. They have converted their collection vehicles to electric power, but more interesting is their technology and method of extracting metal from slag (residue from incineration). This technology has led to a truly circular business model because they are now able to recycle a rare earth species as cutting sand used in industrial processes. They take the used sand back from customers, clean it, and add a share of virgin sand, after which it is resold. The mixed sand has shown better properties than the virgin sand and it is a good business case for both Meldgaard and their customers to recycle because customers do not have to pay for sending their residual sand to expensive deposits. Now the business model has extended to also include old glass fractions that are not directly recyclable as glass and can be recycled here as sand for industrial processes. This case is also in the Case Collection (Haar, Nordic Case Collection, 2024b/24 (to be published)). In the case collection, there are also other good recycling cases, such as KALK, which recycles old lime mortar in the production of new ones.

Streams of Materials and Scarcity of Resources

When the legal framework for a Circular Economy, material infrastructures, and resource management in EU is implemented then a sufficient volume can be created of recycled materials. When a reliable volume is available investments in new technologies, facilities, collection methods, business models, and infrastructure will follow. A large part of the green transition may be financed by private companies and private investors. Many politicians believe that the green transition should be financed through taxation, but most new solutions lie in the companies and are profitable if the right market conditions and infrastructure of materials are created. Taxation should rather be used to create incentives to do differently.

It is difficult to clearly categorize materials for the new business models because the materials will be used in cascades. First products are reused 1:1 (for example, building materials), then materials can be recycled like wood, steel, mortar, and glass, and finally the totally worn-out materials can be chemically recycled, and biorefined. Chemical recycling is when fibers, molecules, and polymers are used in a chemical process to regenerate a new secondary raw material. Biorefining is part of the chemical recycling where organic molecules can be refined for new biobased materials, such as plastics, fibers, and so forth. The recycling of non-biological (non-organic) materials as metals and glass can live in circular loops almost forever when handled correctly. Remember that the circular material loops must be divided in two: the Biological Circle and the Technical Circle, as illustrated in the butterfly

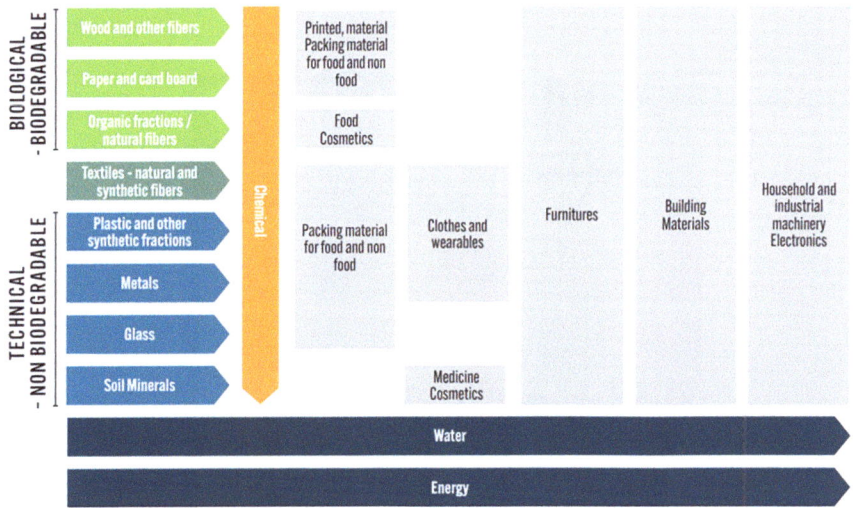

Fig. 9.8 Material streams to products

by Ellen McArthur, also described earlier in this book. Fig. 9.8 is an overview of material types and the flow of products they are part of.

Many analyses have been made on future demand for raw materials. The transition to RE requires many of the raw materials, especially metals, that are becoming scarce, and access to these materials is seen to be a significant barrier in the green transition. Providing for 10 billion people in the future in a linear economy will soon bring an end to access to many critical resources. Figure 9.9 shows a representation of flows of raw materials in nine selected technologies, 3 sectors, and 25 raw materials presented by the EUC. These three sectors are dependent on many of the scarce resources and critical for the transition to a clean economy based on RE.

The large recycling company Remondis SE & Co. has also made some interesting estimates of the availability and supply of selected raw materials. This shows that we are already experiencing scarcity. The estimates also state the poor recycling rates of these raw materials and the potential for substitution to other raw materials. No doubt, we must bring an end to the linear economy. Investigate the estimate of Remondis on: https://www.recyclingrohstoffe.de/fileadmin/user_upload/Recyclingrohstoffe/Infografik_Rohstoffknappheit_DE.pdf

EU has funded a project called FutuRaM, that aims to change this by introducing a strategy, reporting structure, and guidelines for raw materials' use, which will massively improve our data and knowledge on those materials. The findings will assist in assessing waste stream recoverability and the recycling of secondary and critical raw materials, significant for a green future (https://cordis.europa.eu/project/id/101058522).

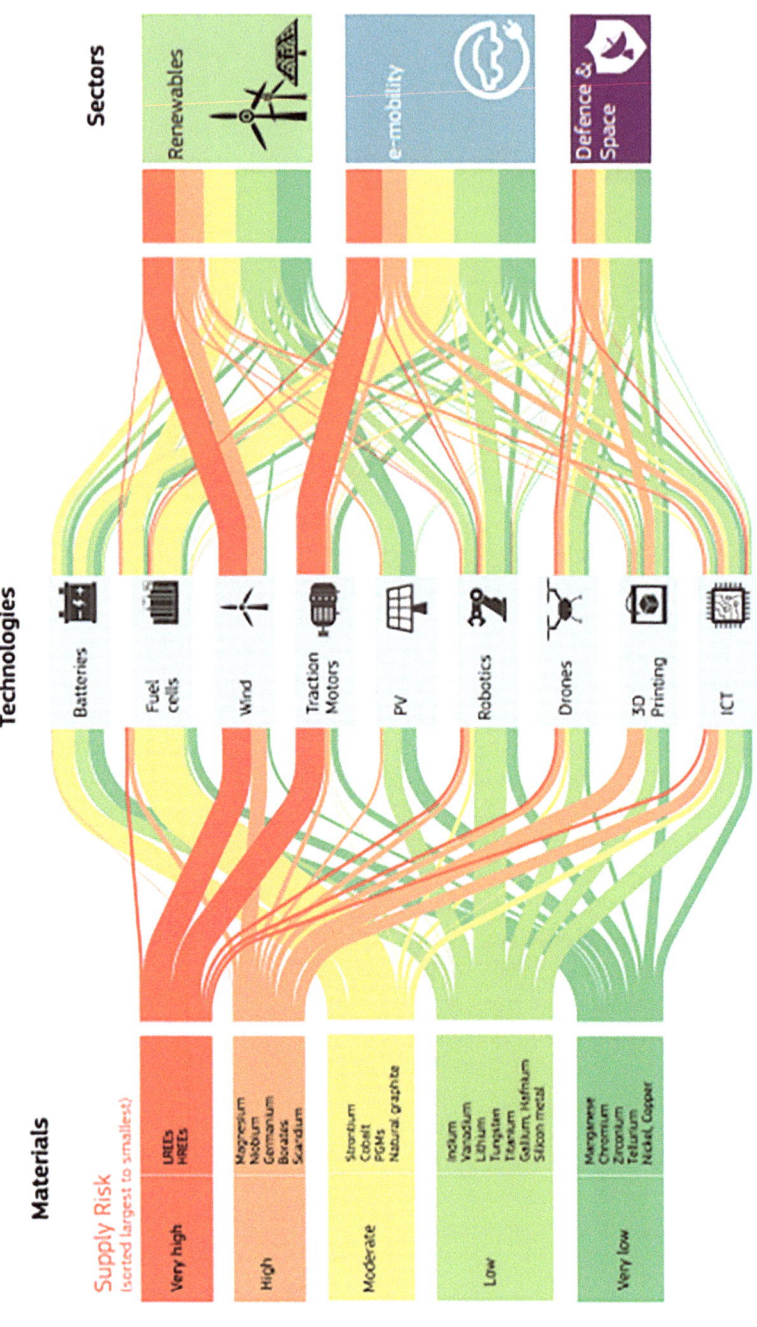

Fig. 9.9 Representation of flows of raw materials and current supply risk. Semi-quantitative representation of flows of raw materials and their current supply risk to the nine selected technologies and three sectors (based on 25 selected materials). https://rmis.jrc.ec.europa.eu/uploads/CRMs_for_Strategic_Technologies_and_Sectors_in_the_EU_2020.pdf

Supply of Recycled Materials

The new material banks and the recycling industry must build critical volume, credibility, and security of the supply of recycled materials. Therefore, they will become responsible for:

- Collection of materials and worn-out products.
- Sorting the collected materials into uniform and standardized fractions according to sorting standards.
- Description and categorization of materials and their substances, including chemical contents.
- Achieving formal approval and quality assurance for resale by certification schemes.
- Labelling and standardization of fractions and processes—safety, chemical, product, and eco-labelling.
- Display, sale, and distribution of materials and semi-products.
- Advice for producers on circular product designs for disassembly.
- Offering take-back systems.

The Circular Economy is to replace the present supply of virgin materials with reused and recycled materials. Therefore, it is central to be able to offer a sufficient volume of each uniform reused and recycled material to manufacturers to the same extent as the new ones today. Only when producers find security in the supply of recycled materials, they invest in new circular business models and the technology to scale the Circular Economy.

> **Trust must be created for producers to be able to:**
> - Purchase sufficiently uniform quality volumes of reused and recycled materials.
> - Purchase recycled materials free from harmful chemicals, even of older age. Also, from products where chemicals may occur that today are not permitted.
> - Achieve guarantees that reused and recycled materials are indeed recycled materials or products to comply with the EU legislation and the material loops available, as well as availability of infrastructure.
> - Mitigate risk from any potential problems in future arising from the use of recycled or reused materials—potentially done by a new type of insurance.

The manufacturers will require guarantees on the same level as for virgin materials. This means that the recycling industry or new actors must find new ways to provide these guarantees. This creates new business opportunities for producers, recyclers, and insurance companies to support the new material loops of recycled materials. Some "first movers" are expected to invest in technology and strategic collaborations early on to increase the material volumes, thus creating new take-back systems and the access the recycled materials first.

In countries with no or hardly any waste collection or waste management, waste incineration still makes sense to clean up the landfills and waste in the natural environment. In those cases, it makes sense to use waste for energy production, also because the waste in landfills often does not represent a resource value any longer. Incineration of waste displaces fossil fuel, although it is not climate neutral, as widely believed. However, incineration will typically be more climate friendly than the coal still used by many countries for energy production. And it is part of a large-scale clean-up that will help avoiding discharge of environmentally dangerous substances and chemicals, and plastic into the natural environments. Thus, it requires environmentally friendly facilities and goals of switching to resource efficiency instead of waste production. Countries, such as Denmark, that today depend on energy from the incineration of waste are challenged in the transition to Circular Economy because it is also a transition of the energy system.

Chemistry in the Circular Economy

Since the 1960s, the development of chemical additives for many types of products has been ongoing. Many of these chemicals now turn out to have damaging effects on people and the natural environments. Today, we are surrounded by products that emit harmful chemical substances when used—this applies to building materials, furniture, and textiles. The EU's chemicals legislation REACH, is the backbone of European chemicals legislation, has shifted the burden of proof onto the manufacturer, and a completely different restrictive and cautious approach to the use of chemicals in the future.

REACH operates with positive and negative lists of chemicals and additives that are continuously updated. The basis for regulating against harmful substances already exists in the legislation. Now the challenge is to identify and minimize the harmfulness of many chemicals which will be banned and replaced by much less harmful substances. The EU has developed a new reporting system for critical ingredients, called the SCIP database (Substances of Concern in Products). Reporting requirements have entered into force and companies must report if they have products containing more than 0.1% of critically listed substances. You can learn more about SCIP at ECHA (European Chemical Agency—https://echa.europa.eu/scip).

A challenge in the recycling of materials is contamination from chemical additives, which can make recycling difficult. Over decades and up to the present days, chemical substances have often been added to new products to create new properties, so that the materials can be used in new ways or products can be manufactured cheaper. Chemistry is also important in achieving different product properties, but historically there has not been sufficient focus on the long-term, harmful effects of chemical additives. Now, it turns out that some of these chemicals are damaging to people and the natural environments. Chemicals are polluting many materials and material flows, and the harmful substances are unfortunately concentrated when the materials are recycled. Problems with controlling and removing the added substances in the recycling processes are emerging and a barrier for the Circular Economy.

> Chemical pollution does not directly relate to the Circular Economy, but when the materials are recycled, the chemicals are concentrated and mixed further. Thereby accelerating the cocktail effect (mixing different types of chemicals) of these chemicals. This creates a bigger challenge in the recycled than in the original materials. Creating clean material loops benefit the Circular Economy, health and the environment.

An example of this challenge from chemicals is the recycling of cardboard and paper. There are good recycling systems of wooden fibers in EU, but ink, varnish, glue, and other chemicals are added to the paper and cardboard, so the potential for recycling is limited. Despite removal of the harmful chemicals in the recycling process the recycled materials contain harmful chemicals and cannot be used for direct food contact due to risks of migration of the harmful substance to the food.

Food contact materials (FCM) hold a special challenge here. In the EU, it remains the responsibility of the company to ensure that the materials are suitable for food contact. This means that food today is found in contact with recycled paper and cardboard, and inks contain harmful chemicals that may migrate to the food. In other countries such as Switzerland, the USA, and Canada, there is strict legislation on this. Methods are available to uncover if harmful chemicals migrate to the food (MOSH/MOAH). Thus, there is not a systematic phase-out of these critical elements.

Other chemical compounds that have been shown to be particularly harmful to human health are fluorinated substances (PFC, PFAS, perfluorinated substances, fluorocarbons, etc.), which cover a large group of chemicals that provide water, dirt, and grease-repellent properties and other properties. They are found in our clothes, pots and pans, food packaging, and many other places. They probably turn out to be significantly more harmful to humans than first realized. They accumulate in nature because of discharges from production sites, and do not decompose. They are now found in soils, water streams and grasslands in Denmark, and probably all over the world. They are also found in human milk fed to babies accumulated from our food, as meat and fish.

In future we will experience chemical scandals occurred by accumulation of chemicals in our food chains due to the lack of systematic ban from the EU of groups of chemical compounds with harmful properties. Roughly speaking, now the EU prohibits one substance at a time. A lot of money and research is spent on phasing out and replacing individual substances. Without further hedging and holistic identification of whether the new substance also carries (other) harmful effects. The same is true for pesticides, which now show health-damaging effects and migration through the soil and into drinking water. This went undetected when the substances were introduced. This new knowledge of the challenges hopefully will cause a different approach from the EU to phase-out various chemical groups. It must be expected that this means banning a lot of chemical substances as the healthcare consequences of these hazardous chemicals are uncovered.

Companies are currently focusing more on the problems with chemicals because consumers have increased suspicion of the ingredients and chemical contamination. The consumer does not know what is in the products and how they are manufactured. This is particularly a pronounced problem in Asia, where many scandals have been revealed involving added substances, especially in food. But an increase in chemical scandals is also to be expected in the EU, since we also here have been generous with the use of chemicals. This problem relates to almost all product categories and our foods and is very difficult to get the full picture of.

Examples of Circular Business Models with New Material Loops

The Circular Economy requires new business models to create loops of materials. The business models must be specific for each type of product and need to interact with the new material loops for:

- Building materials
- Electronics
- Textiles
- Plastic
- Paper and cardboard
- Glass
- Metals
- Organic fractions

It is important to note that now it is often cheaper to mine recycled materials from landfills than virgin materials from mines. This welcomes upcycling and recycling as a financial model and is a precondition for new business models and new material loops based on recycled materials.

Packaging materials, plastic, textiles, and building materials are some of the products and materials that are challenged by a lack of circular infrastructure and low rates of recycling. Examples of new material loops and business models for the most important materials are included in this chapter to illustrate the challenges in the transition to a Circular Economy. Plastic, paper, and cardboard are important packaging materials, but also important materials in other products. Textiles are included to show the challenges of the long global value chains. Finally, building materials are included as another example of a significant waste fraction that requires completely new infrastructure, new material loops, and new business models all along the value chains to meet the Circular Economy.

Packing Material: Plastic, Paper, and Cardboard

Packaging materials make up a large part of household waste and are a major concern for consumers because they experience large amounts of single-use plastics and disposable packaging in their bins that are unsustainable every day. In general, it is challenging to find out which types of packaging materials are sustainable, and which are recycled. Packaging is made from different types of materials that have very different levels of recycling, ranging from plastics, where there is currently very little recycling, to glass and metals, where good systems exist (Haar, White Paper on Packaging in a Circular Economy, 2021).

Below, plastic as well as fiber and cardboard are reviewed as examples of materials where infrastructure and business models need to be changed in a Circular Economy. It will affect the amount of packaging but also offer new material loops for products other than packaging. New material loops for packaging are a good example of how new material infrastructure overlaps with other products and helps to create new paths in the Circular Economy.

Plastic

The production and consumption of plastic emits 400 million tons of CO_2 annually, and a significant share of this plastic ends up in nature. Plastic is a problematic material, especially when used for packaging. Up to 90% of the CO_2 emitted by the production and burning of plastic today can be saved by full reuse and recycling of plastic. Plastic has a great potential to become a sustainable material, because it offers good durability for food, and holds the properties that are demanded for. Nevertheless, plastic creates major challenges because there is so much plastic on the planet already, it is mixed and often not suitable for recycling. Chemical recycling, where mixed, used plastic is broken down into oil (polymers) and regenerated as plastic, is considered recycling by the EU. There is really no need to produce more virgin plastic, and scalable recycling of plastic must start now.

Plastic covers many different types of materials used in many different products, and packaging accounts for 40% of all plastic produced. Detailed information on plastics can be found on the website of the European Plastics Industry website (https://plasticseurope.org/) and in their design guide (Signers, 2020). More detailed information on packaging materials, legislation, and sustainable packaging can be found in the White Paper on Packaging materials in a Circular Economy.

Selected EU countries and companies in the plastics industry have signed the European Plastics Pact, which leads the way to a circular infrastructure on plastics, see Figs. 9.10. 51.2 million tons of plastic are produced in the EU, and 1.5 million tons are exported out of the EU (Signers, 2020) as waste and disposed in, for example, Asia. Lack of ensuring proper disposal in the value chains is why plastic packaging from Europe ends up in rivers in Asia.

At the bottom of Fig. 9.10, an overall circular infrastructure for the reuse and recycling of plastics is proposed. In the cogwheels, a circular infrastructure is outlined as the need for harmonization, cooperation, and connection of material loops. The circular infrastructure must create a sufficient volume of all types of recycled plastics, especially for PET, which is the main type used for food packing. There is a lack of recycling systems and recycled materials in the market, which is reflected in the prices of rPET (recycled PET), being higher than the price of virgin plastic.

Establishing the circular infrastructure requires investments and fees to cover the handling costs. This means that the price of plastic will rise to finance this, and an

Fig. 9.10 European Plastic Pact. Source: https://europeanplasticspact.org/

end to just duping it free of charge. The EU has adopted a Single-Use Plastics Directive and an Extended Producer Responsibility (EPR) on packaging materials. This affects the manufacturers of the packed products. This EPR has effect as of

January 2025, and taxation is one of the elements in achieving the target for recycling plastics (EC, Circular Economy, definition, importance and benefits, 2015).

The establishment of a circular infrastructure for plastics and packaging materials is at the forefront of the roll out of the EU's CE Action Plan, and there is a strong focus on this transition now. Member countries will have to implement local legislation and structure for packaging and plastic handling by 2024. Then plastic has great potential as a sustainable material. When reuse and recycling are fully implemented and virgin plastic is no longer produced, plastic will be just as sustainable as other material as metals and glass, or even more. And it holds some particularly good properties for food packaging and to counter food waste.

Paper and Cardboard

Figure 9.11 shows a new business model on fiber-based materials inspired by KLS Pureprint—a Danish packing and printing company that works actively on transforming the waste stream of compostable fiber-based materials. First, it is necessary to create clean materials and products, both for paper and cardboard, especially in additives as, ink, varnish, and coating. Then transparency and traceability are necessary on all the ingredients added to the printed materials and packing materials.

It is a challenge in the printing industry that companies through the value chain cannot obtain knowledge of chemical content. This is why composting and utilization of the nutrients from the degraded printed matter are not allowed today. Once critical ingredients and chemicals have been removed from the printed matter and the packaging materials, clean material streams can be created, resulting in both a cleaner recycling stream of wooden fibers for materials and new clean material loops for composting some of the packaging materials. Contamination of printed materials is really a shame because the EU has a good system for recycling these natural fibers. The addition of chemical compounds limits the possibilities of recycling and increases the cleaning cost in the recycling process.

Cardboard and paper that have been in contact with wet or oily food, are banned from entering the recycling and a significant proportion of wooden fibers cannot be recycled. So, it is important to have a stream for compostable materials where the nutrients can be preserved and utilized on the agricultural soils and energy can be utilized from biogasing. The establishment of clean material streams with less chemistry is slowly moving for fiber-based materials because now consumers become more focused on the chemistry and the composting of natural materials. This will create more sustainable footprints in each step of the new streams as well as raise the quality of the recycled products and the overall footprint of fiber-based materials.

Textiles

The chemical aspect described above is true also for several other material fractions including textiles. We just do not have a recycling system for textiles in the EU, which enables us to see this. Heightened awareness and stricter legislation are needed to make clean material streams provide maximal value from recycling with no damage to the environment or people.

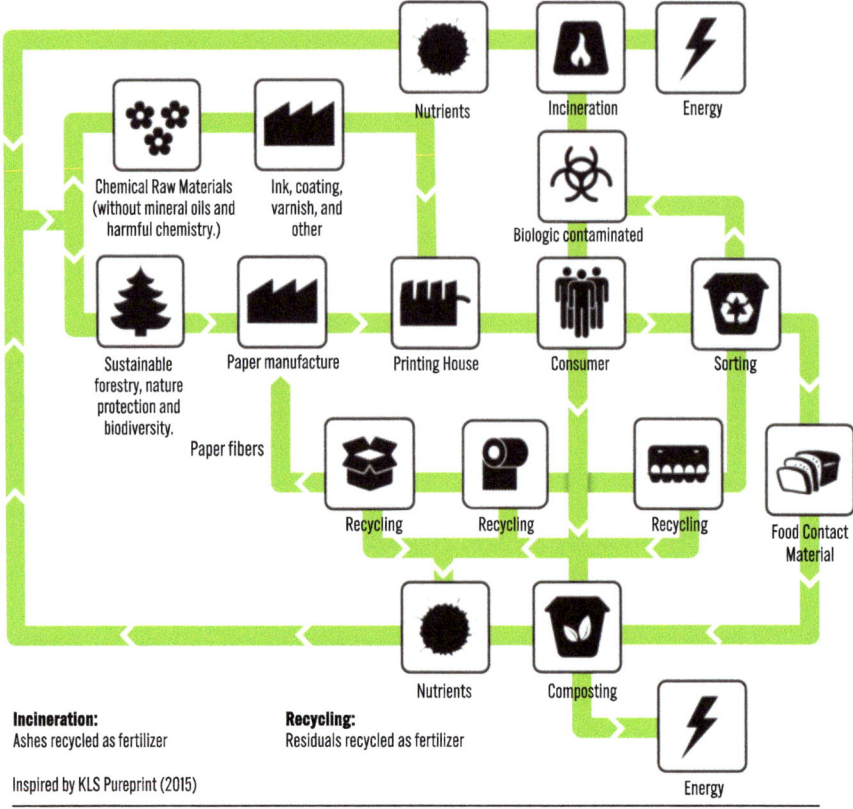

New green business model for fiber-based materials as paper and cardboard with better use of the raw material in recycling due to clean streams and because the residuals can be composted and used as nutrients on farm soil.

Fig. 9.11 New Business Model for Fiber-based paper and cardboard. New green business model for fiber-based materials as paper and cardboard with better use of the raw material in recycling due to clean streams and because the residuals can be composted and used as nutrients on farm soil

Figure 9.12 gives an overview of the many business models and materials streams of the textile industry. This represents the most challenging global value chains where consumers have been completely decoupled from understanding the production of textiles and clothes and their environmental and social impacts.

> **The material stream of textiles consists of at least four major sub-business models.**
> - Primary production of natural and synthetic fibers—two different materials that are mixed in the textiles (Asia).
> - Secondary production of garments and clothes (Asia).
> - Design, Marketing, and Consumption (Western World) with a very little share of reuse here.
> - Waste where a large percentage is still sent to landfill or burning, and a very small fraction is sent back for recycling and re-spinning of fibers.

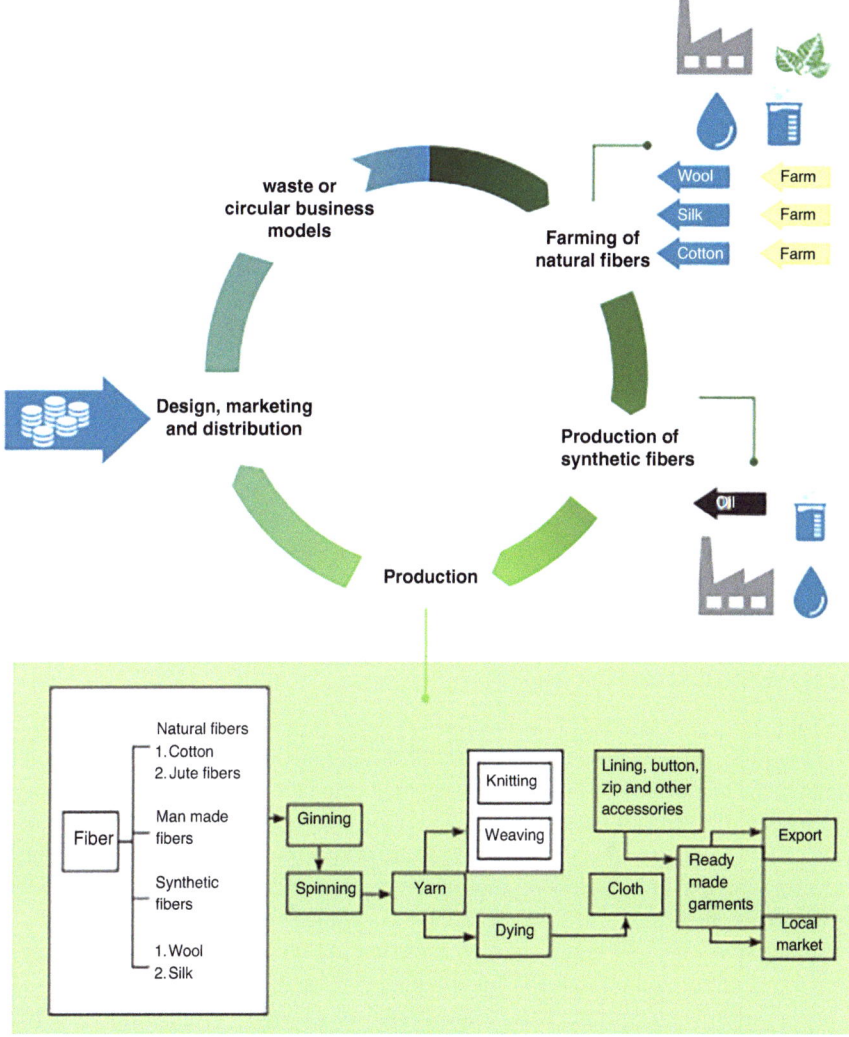

Fig. 9.12 Textile value chains

These material streams in the textile industry illustrate the falls of the linear business models and the model of "take-make-waste" resulting in very low prices on textiles and enormous environmental impacts and social biases largely outside the consciousness of consumers. In the textile industry, there is a huge overproduction of product series to meet demand and overconsumption. This is only possible due to the very low production costs. Clothes are produced for the fashion industry where only a share of the produced goods is sold, the rest is wasted or burned. The production prices are so low that retailers rather risk stocking up clothes than running out of stock on their shelves.

Brands destroy large quantities of clothing because they must maintain an exclusive image and cannot allow the on-sale prices of their clothes. Some of the American and European large retailers are getting away with this unhealthy business model without taking genuine care of the primary or secondary production in Asia. One large challenge in the fashion industry is a heavy overproduction and overconsumption of clothes without any responsibility for the externalities and the impacts.

The textile industry has other large challenges regarding their social impacts, as labor rights, at the production sites, and the high pollution from pesticides, dyes, and water consumption. And to this is added climate impacts accounting for up to 10% of the total GHG emission globally. This industry really needs transformation on all levels from primary production of natural fibers and plastic fibers, to manufacturing processes and consumer business models. This industry is only able to hold on to linear business models in fast fashion because consumers do not have transparent and traceable data available. Pricing is one thing that must change here because it is simply too cheap to produce clothes, and the profits are only at the late end of the value chains with the marketeers.

The fashion industry is driven by its ability to create consumer trends and motivate rapid changes in demand. This became obvious during the Corona pandemic, where some of the large European brands as Bestseller and H&M lost large amounts of revenue and share values dropped because they ended up with stock of clothes waiting to be run over by the next bimonthly fast-fashion release. The solution here is to create transparency and traceability followed by due diligence of products and taxation to counter the linear overconsumption, waste of resources, and violation of human rights and the environment.

Buildings

The construction industry is the industry that generates the most waste, and unfortunately the linear business models have a good grip here. Today, a lot of old, good-quality building materials are thrown out and replaced with new ones holding a much shorter lifespan than the old ones. There is enormous potential in reusing and recycling the existing instead of continuous production of new materials at cheap prices. This also means renovating rather than building new. There is already a strong focus on the green and circular transition in this industry, but there is still a long way before circular loops, infrastructure, and business models are in place. A circular business value chain for the construction industry is outlined in Fig. 9.13.

The existing buildings will become the future material banks, supplying building materials in the Circular Economy. The existing building stock must be mapped as soon as possible and all the building materials here must be registered as the basis for future circular constructions. In the future circular loops include repair and reuse of windows, doors, bricks, boards, beams, and so on to be reinstalled in new or renovated buildings. Later materials can be recycled into new building materials or other products. There are already asset owners who require recycled and reused materials in the new buildings and require that materials must also be reusable and recyclable in the future as well. Now, the market cannot meet this demand for reused and

Examples of Circular Business Models with New Material Loops

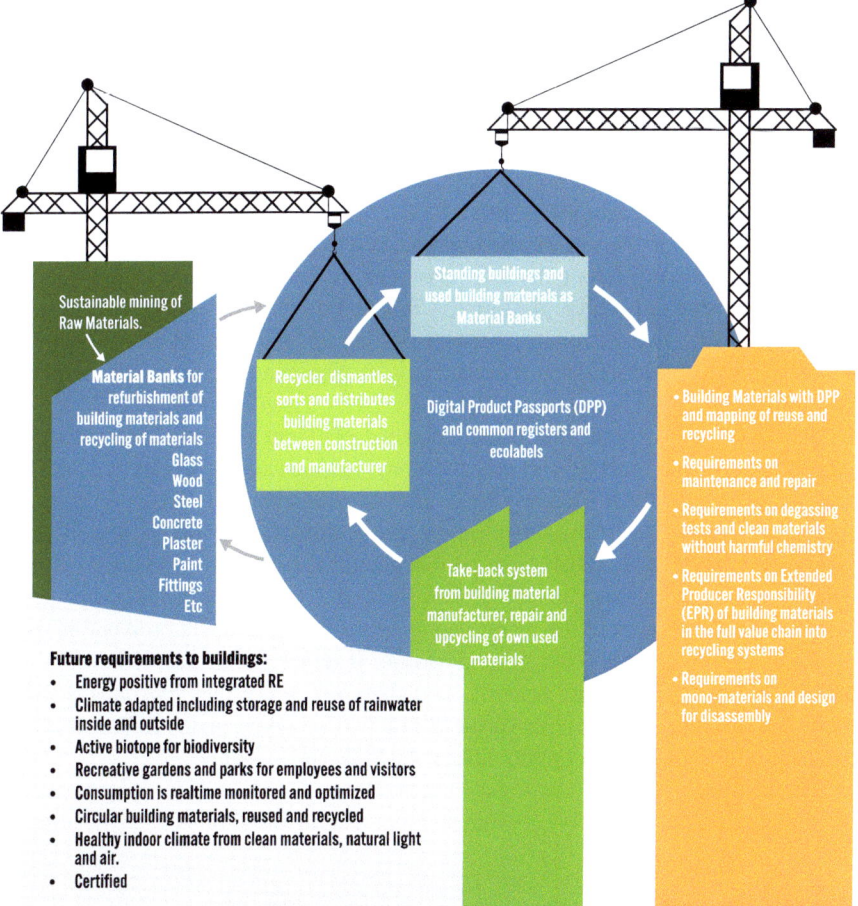

Fig. 9.13 Circular value chains in construction

recycled materials and scalable business models are needed. The mapping of the existing building materials in buildings is taking place in Holland and in Denmark (Circle), and possibly also in many other countries.

Building materials and products will carry a digital product passport (DPP) in the future, and new eco-design criteria are on the way from the EU. This will allow all stakeholders in the value chain to have access to information on the materials, as described in Chap. 6. The DPP is the key element in making this industry circular, and when implemented in full scale, the need for virgin materials and new building materials will be drastically reduced.

In part III of this book, a guideline on sustainable buildings is provided.

Sharing Economy Versus Circular Economy

Many people confuse Sharing Economy and Circular Economy. Here, it is explained how they differ and how they overlap. Both the Circular Economy and the Sharing Economy challenge and disrupt traditional business models and societal structures. Perhaps this is a reason for the confusion. The overlap can be illustrated in Fig. 9.14.

The goal of the Circular Economy is to ensure access to and use of material resources through new ways of return, reuse, and recycling of products and materials. Whereas Sharing Economy is a consumer model that creates a new kind of consumption through digital platforms. Circular Economy is almost always more sustainable than a linear economy because reusing and recycling resources removes waste and the production of new products with significantly lower environmental and climate impacts.

The Sharing Economy, and in particular the platform economy, promotes consumption and is a competitor to the traditional linear delivery model. It is not the purpose of the platform economy to be more sustainable or to focus on reuse or recycle of resources. It aims to promote consumption, new services, and often increases consumption. When more people get cheaper access to products without owning the consumption increases. However, the Sharing Economy may save resources in those parts of the world where overconsumption is a problem because products are not fully utilized to its potential. It requires that people be aware of environmental and resource scarcity, and therefore reuse and recycle resources using the Sharing Economy platforms. Table 9.1 list some of the differences and elements of the Circular Economy versus the Sharing Economy.

The Sharing Economy may reduce the need for overproduction and purchasing of new equipment by every consumer. After all, in most developed economies

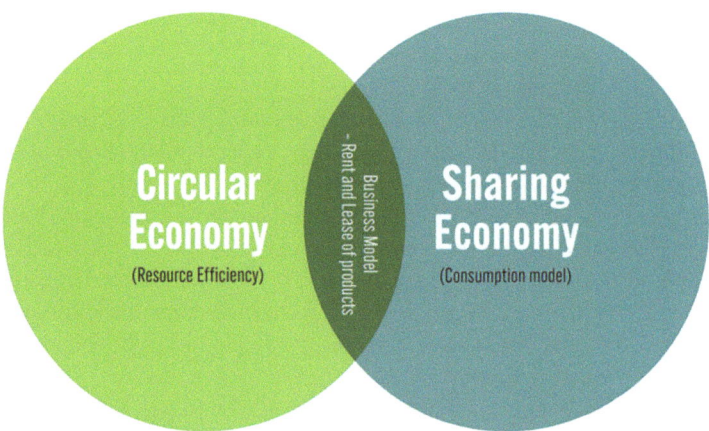

Fig. 9.14 Circular Economy vs. Sharing Economy

Table 9.1 Elements in Circular Economy and Sharing Economy

	Circular economy	Sharing economy
Goal	Avoid exploitation of virgin resources and hazardous mining of natural environments as well as securing access to scarce resources.	Creating new consumer and delivery models and involving new consumers and suppliers.
	Secure resource efficiency. Strategically important for the way we design and produce. Partial impact on our supply models.	Advances consumption through new online platforms. Strategically important for the way trade is executed and for the communication between supplier and consumer. Only limited effect on product design and production.
Growth	Advances growth through mobilizing values otherwise wasted.	Advances growth through increased consumption and a more efficient use of human resources.
Business model	Creates new products and take-back systems through business models.	Creates new business models between new types of consumers and suppliers.
Impact	A redesign of our production and economic models—a new industrial revolution.	A consumption model—a service revolution, and an ICT revolution (Industry 4.0).
Legislation	Implementation of EU Strategy and Action Plan on Circular Economy and new national regulation for creating infrastructure for reuse of products and recycling of materials.	Requires new legislation for taxation, control, and insurance of supply (services).
Spread	Predominantly powered B2B involving consumers.	Predominantly C2C and B2C, and to less extent B2B.
	Product-driven	**Supply-driven**

people overconsume and sharing economy may regulate this. However, in developing economies we may see an increase in consumption, Sharing Economy may give a new type of purchase power.

Like Circular Economy, the Sharing Economy is an old phenomenon, covering a system where the consumers share products and services. Instead of everybody owning a product, products are shared, i.e., borrow, or lease products from each other. This results in rendering services instead of having ownership. Sharing Economy is boosting these years because now online-based platforms give far more access to connecting—consumer to consumer as well as business to consumer. The unique thing about Sharing Economy is that new businesses are born, offering new kinds of services. A standardization of the services offered for example by Airbnb and Uber is established, providing a rating system and transparency that is useful for the consumer. Sharing Economy challenges the traditional structures of how to supply, and how to control supplies and the quality of products and services. Above all, it is a challenge to the taxation of business transactions.

> **Sharing Economy is given much attention because:**
> - It creates a new kind of consumption, which is far more consumer-driven than traditional trade, which is largely producer or supplier-driven. For decades, the companies have tried to implement consumer-driven innovation, because the perception has been that this facilitates rethinking and new business opportunities. The sharing economy supports this.
> - It creates a unique transparency in the transaction between supplier and consumer, which is an advantage to the consumer, and also an advantage to the provider.
> - It challenges the traditional suppliers and suppliers will not always be companies. Owning or having access to a product makes all potential suppliers.
> - Quality control, insurance, consumer protection, and taxation of services are seriously challenged and there is a lack of legislation and control facilities. When the suppliers are everybody and the delivery model is different as in the traditional economy, the government is not able to control the transactions and agreements as they are used to.

Sharing Economy may create a challenge to our society models and our taxation systems in the developed part of the world. While it has an enormous growth potential in the less developed part of the world, where taxes, control, and insurance are not the pivot of the economy and the society models. With the spread of the platform economy, the necessity of switching taxation from work and transactions of goods (deliveries) to physical units, for practical reasons and because it is controllable. Ownership of physical units as houses, land, cars, etc. are much easier controlled than to control who is using it.

Some people believe that the platform economy constitutes a threat to welfare society and consumer safety. Others believe that it is the challenge necessary to vitalize the entrenched welfare system—especially in Europe. All these new kinds of transactions among a far less well-defined crowd of suppliers and consumers will change the need for new taxation systems and securing the consumer in the future. The Sharing Economy may be an incredible opportunity to increase productivity in the very established and developed societies. Here the Sharing Economy can create a larger resource efficiency and a better use of human resources, that are today used to manage sales and delivery. Increased transparency, traceability, easier access, and communication with the consumers, are what make the Sharing Economy contribute with a new thinking.

Platform Economy, Reuse, and Product as a Service (PaaS)
New consumption patterns emerge out of online sharing platforms for houses, cars, clothes, and other devices. At the same time sale of reused furniture, clothes, electronics, building materials, and other household applications is rising. The sale of reused goods has risen in the last 10–15 years mainly because sellers and buyers

easily can find each other online. It also reflects a move from overconsumption and a make-take-waste culture toward reuse. Old products are often of better quality than much new stuff, which intensifies the interest in the sale of reused goods also among people who can easily afford new stuff. Reuse has become a fashionable consumer trend showing humans surplus and responsibility, and retro designs are booming these years. People get more invested in covering a specific need than owning stuff.

New common trends, especially among young consumers are service models competing with the traditional linear consumption models. Making a product available as a service for someone with a need for it for a limited period. Often young people do not show the same need for ownership, or the same financial opportunity compared to previous generations. Probably because the young generations are raised in a culture of abundance and do not fear having enough. People become more invested in covering a need in a period rather than owning stuff. Having the opportunity to sit in a chair regardless of the chair being rented. Transport by car rather than owning a car. This is meaningful in the increasing urbanization, where people live closer and have fewer storage room.

A further consequence of the new service models (non-ownership business models) is that maintenance and repair become the responsibility of the leaser or owner and not the user, hopefully with extended lifetime of the products. This trend is also spreading in B2B, where robots and other technical installations are rented or leased to a greater extent than purchased. This not only provides financing of equipment but often also a service agreement and the possibility of redeploying the machines to other users in case of changing needs.

It is important to distinguish between service models that further the Circular Economy and the recycling of materials, and those that counteract it. Many financiers and providers offer leasing and service contractors to customers to enhance sales in a linear economy, thereby financing a purchase that the consumer would not be able to afford. Financing models of cars are in Denmark used to avoid taxation, especially combustion cars. So, if service models are to belong to the Circular Economy, they must facilitate reuse, maintenance, and recycling of the products. Typically, when the manufacturer retains ownership and offers a bonus or a deposit to take-back the products and keep them in the loops.

As an example of an inappropriate service model that did not promote the reuse or recycling of products, is when Phillips introduced the renting light rather than selling light fittings. The case is described earlier in this book. This case could have been a brilliant model for preserving light fittings and the reuse and recycling of these. Unfortunately, the income from energy savings of this model harvest by Phillips was so attractive, and customers were uncritical of the environmental impacts of the model. So, when this led to the installation of a huge number of new, cheap single-use fittings with LED, that were certainly not circular, it became an acceleration of a linear business model instead of a PaaS model, that everyone looked for.

Climate Nexus on Circular Economy

The transition to a Circular Economy is sustainable because it is:

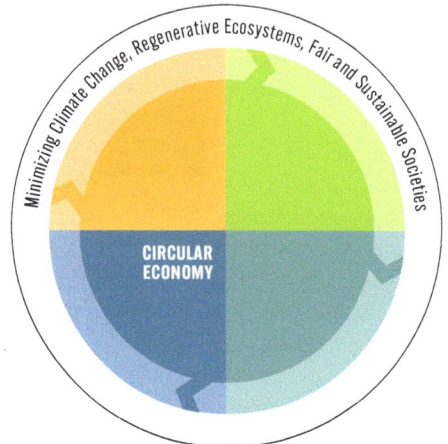

Restoring scarce resources by redesigning, reusing, and recycling.

Counteracting environmental disasters seen globally due to tremendous amounts of waste being exported from the consuming countries and deposited in the developing countries, resulting in deposits of waste as plastic, textiles, electronics with the result of pollution and other environmental disasters.

Replacing products and quality designed for the linear economy with products that are designed for the circular economy, thus increasing the potential for a sustainable lifestyle of future generations.

Controlling and replacing chemical pollutants from material streams and pollutants introduced earlier without accounting for the healthcare and environmental problems they caused.

Creating immediate energy savings from recycling compared to extraction of virgin resources.

Creating regional value chains that are transparent and traceable to ensure sustainable consumption to stabilize and develop regional economies and wealth.

The basis for regenerative ecosystems due to the decoupling of land use for mining of virgin resources.

Redesigning the global value chains to regain control of economy, and security of supply.

References

EC. (2015). *Circular Economy, definition, importance and benefits*. https://www.europarl.europa.eu/news/en/headlines/economy/2015120STO05603/circular-economy_definition_importance-and-benefits:EU Commission

Haar, G. (2021). *White Paper on Packaing in a Circular Economy*. Cph: KLS Pureprint.

Haar, G. (2024a). Rethink Economics and Business Models. *Rethink Economics*. SpringerNature.

Haar, G. (2024b). *Nordic Case Collection*. SpringerNature.

Signers, E. (2020). *European Plastics Acts*. wrap.org.uk.

Chapter 10
Transition to Sustainable Public and Individual Transport

In the transportation sector, road transport is absolutely the largest emitter of GHG, and the largest growing part as well. See Fig. 10.1. Road transport covers transportation of people over land, and an increasing proportion of land freight of goods, both switched from rail and ships. Transport by train is almost unchanged over the past 40 years, and the increase in shipping has not followed the overall increase in emissions from transport. Trains and shipping are absolutely the most energy effective modes of transportation and for the past decades they have been further streamlined and more energy efficient, as trains have become electrified, and ships become larger and more efficient in energy consumption.

The technology is available to remove particulate pollution from shipping, which is a major environmental problem. The industry constantly claims that this must be regulated internationally not to create uneven competition, but now sustainability has become a competitive advantage, so operators just need to get started—both on board the ships and in the ports, where the ships dock. Shipping creates a large environmental impact by pollution of the air in and around the cities. The Danish company Maersk, one of the largest shipping operators globally, just released a profit of 29.3 billion US$ for the year 2022, and that requires responsibility in implementing the technologies to avoid air pollution. Maersk has a Net Zero emission target by 2040 and avoiding air pollution must be on the same top priority, at least. With this kind of profit and with the proven competitive advantage of genuine sustainability it is not a question of, it is a question when to deliver on… -"a **societal commitment** to act and drive impact in this decade", as Maersk state in their Roadmap to deliver net zero by 2040.

With just over 10%, aviation represents a small share of the total climate impact from transport in a global perspective, but still receives a lot of attention. Emissions are rising rapidly from international aviation and have doubled since the turn of the millennium. However, aviation represents a large share of the climate impact on the people or goods transported, and in that respect, with increasing prosperity and

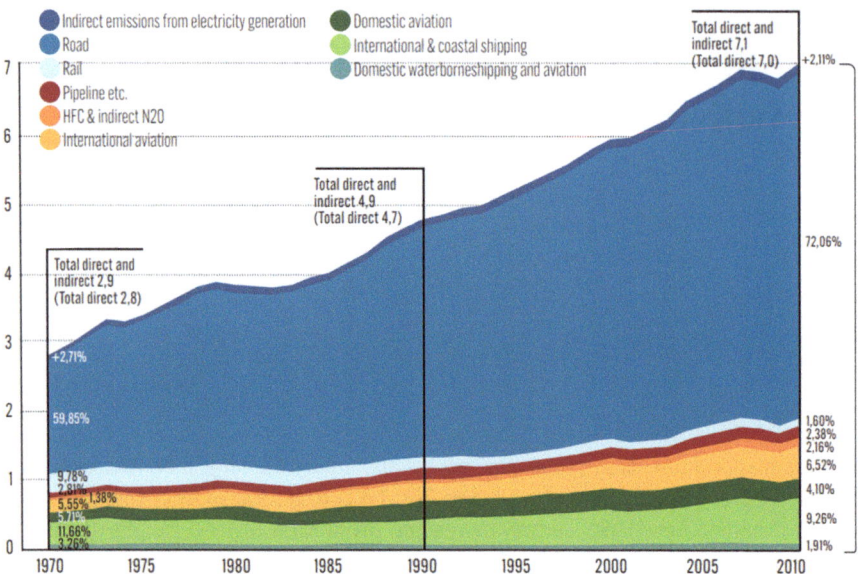

Fig. 10.1 **GHG by means of transport**. Direct GHG emissions of the transport sector shown here by sector mode rose 250% from 2.8 $GtCO_{2eq}$ worldwide in 1970 to 7.0 $GtCO_{2eq}$ in 2010. (IEA, 2012a; IRC/PBL, 2013)

population growth, it is necessary that aviation becomes climate neutral, as soon as possible.

In addition to climate impacts, fossil fueled transport creates several challenges such as:

- supporting an inefficient, linear technology,
- extensive pollution, especially in cities from both particle and noise pollution, and,
- individual transport causes congestion, traffic jam, stress, and inefficient travel time. As a result of the transition away from historically electrified, public transport, such as trams and trains, towards fossil-fueled individual transport.

A large share of our transport today is powered by fossil fuel internal combustion engines (ICE). The combustion technology was a very inexpedient technology that unfortunately won the race over the electric engine a 100 years ago. The electric engine was invented at approximately the same time as the internal combustion engine in the second half of the 1800s in the early years of industrialization. Maybe the combustion engine became the start of the linear economy with a linear input of a resource causing emission of GHG and pollution. The ICE utilizes only a small proportion of the energy for propulsion, while a large part of the energy is wasted as heat. It is estimated that between 50% and 60% of the energy in an ICE car is wasted from extraction to propulsion (LCA). This inefficiency results in almost double the energy consumption to move a vehicle as by an electric engine. The electrical engine has an efficiency of as much as 80% from power generation to vehicle (LCA).

Electrified Transport

The electric engine is suitable in a society based on renewable energy. Renewable energy is most often electricity supply, and the electric engine runs on electricity from the batteries storing power from all energy sources. The electricity can be transshipped and stored in the same format, i.e. an efficient system. Cars are already prepared to resupply power back to the electrical grids and are therefore an important storage for overproduction of electricity in the future, for example from wind power at night.

Electric vehicle technology can be combined with hydrogen as a store of power, especially in those parts of the transport sector where batteries still is a challenge. Hydrogen storage is still an immature technology, but is predicted great potential, especially in larger vessels, such as trucks, airplanes, and others. Cars based on supply from hydrogen are also powered by an electric engine, supplemented by an additional storage of hydrogen. The hydrogen technology can quickly be implemented when it is ready and profitable. Hydrogen fueling stations already exist as part of development of the hydrogen cars. The hydrogen storage technology is one of the innovative technologies subject to a lot of R&D in its capacity as Power2X, i.e., the conversion of electricity generated from RE into other non-electrical storage, such as hydrogen.

A shift to renewable energy away from ICE and fossil fuels will almost abolish the particle pollution that today is harming thousands of people with cancer, respiratory diseases, and allergies. Renewable energy will do away with dust and dirt that pollutes our buildings, street, and natural environments. Historically, there was only little concern of the environmental impacts of burning fossil fuels, even though the pollution from old-fashioned diesel combustion was enormous.

A good example of new technologies and digitalization that will promote the green transition is self-driving, driverless vehicles. It will become available through digital platforms to order transportation for a much more accurate end-to-end route planning with an enormous potential for carpooling. When this becomes a reality the number of cars required to support transportation needs will decrease. Estimates shows that only about a quarter of the cars on the roads today will be necessary to provide a better service in a much more sustainable way with the self-driving vehicles. Therefore, there will be a radical shift in the way people are transported and we must expect the same for future freight of goods, services, and information.

Public Transport

Individual transport by vehicles has caused over-traffic and overproduction of cars. We have eliminated some very efficient and sustainable means of transport in the last decades due to the individual means of transportation of people and goods. In many old cities electric trams are still functioning (San Francisco, Lisbon, a.o.). In Copenhagen we removed the electrical trams in the 1970s and even most city-buses still run on diesel with several unsuccessful attempts to go electrical. We need to revitalize transport systems and public transport and re-electrify it. Especially in cities where it is easy to implement, we need to go back to collective means of transportation than individual cars.

More and more cities introduce biking as a new trend and in Denmark, we have a long tradition for biking also providing everyday exercise. A visit to Beijing in 1992 with very wide bicycle-lanes and very narrow roads for cars var. a strong picture. A revisit in 2000 showed the opposite: wide roads and narrow biking lanes. Today Beijing is facing extreme pollution from cars—and it is sad to realize that development can so quickly go in the wrong direction. Hopefully, we will see more mayors as brave as the one in Paris banning cars on selected days with the goal of a full ban of combustion cars in 2030. The solutions to solve the very inefficient traffic in cities goes hand in hand with solving the elimination of GHG emissions.

With a switch to electrified renewable energy, transport can be supplied with sustainable energy. Streamlining and expanding public transport in big cities will create more comfortable and livable urban spaces with room for leisure activities and less stressful transport. At the same time, pollution from combustion can be drastically reduced, just as noise from cars in the city is significantly reduced. Individual transport makes sense outside the cities. Transport must be rethought so that a greater proportion of goods transport is moved away from lorries and towards sustainable and efficient solutions such as rail and sea. Then the distribution of goods in cities can be done with smaller electric vehicles or drones, which both will become driverless, just like self-driving passenger transport. Self-driving technologies offer new opportunities in optimizing transport, and a new semi-industrial mode of public transport may emerge.

Logistics

In addition to the new material loops in a Circular Economy logistics and waste handling are under disruption due to:

- *New circular chains of transportation* disrupting the traditional, linear logistics of our products. The Circular Economy will bring an end to waste, and the products will be provided with new life cycles and new patterns of being moved

around. The goods will be returned to the material bank, producers, etc. More repair and maintenance, and more reuse and retrofitting will influence logistics.
- *More local production* and trade directly to consumers will change the transportation of goods from large-scale economy to smaller units that facilitate the return of packing material—of especially food packing and secondary packing material. The development of these new logistics is at an early stage in the western world, and it is already taking place all over the developing countries, since they never transformed to large scale.
- Extraordinary focus on minimizing single use packing and *new return systems of secondary packing material* will change logistics. Earlier there was a large degree of reuse of packing materials which later years have been replaced with single-use solutions. The new packing directives and guidelines will foster reintroducing of secondary packing for reuse and return.
- New demands from cities all over the world for *cleaner transportation*, such as electrical, hybrid etc. Combustion-driven trucks will be phased out fast because they pollute and make noise and are a considerable source of greenhouse gases. The technology for this transition exists today, and the battery and storage technologies are developing so rapidly that we shall soon see it in freight. These large vehicles provide the space required for batteries.
- *Harbours and trains* are expected to resume their role in the transportation of our goods. Because it is more sustainable and efficient to use this infrastructure than roads. The roads will in future be predominantly used for the transport of people. While the transport of freight and goods will use other infrastructure. Cities will provide new types of infrastructures and will become green, biodiverse, and beautiful to live in.
- *Self-propelled technologies* are soon becoming relevant to transport of goods. The transport can take place at outer times and on long one-way distances between big warehouses outside the cities. Then smaller vehicles will deliver within the cities.
- *Drones* make quite a new mode of transportation, targeting delivery more precisely. Very soon we must expect the deliveries of goods in the cities by drones from big warehouses outside the cities.

Danish and international analyses show that between 30% and 50% of freight trucks drive unladed, meaning that up to 50% of all trucks are empty on the roads. This leaves a large financial potential and room for disruption for those who can reorganize logistics, especially with the changes that both the Circular Economy, digitalisation and new technologies provide. Distribution according to the traditional business models is a financially challenged business with declining markets and declining employment, especially after the Corona crisis. Therefore, new technologies and new business models meeting the Circular Economy will become innovation platforms that can make this industry profitable again. This disruption will come from first movers seeing an opening to get into an industry under pressure. The disruption will also come from large corporations like Amazon, with an understanding of the Circular Economy and with access to the digital technologies, and a fast implementation of new circular business models at their customers.

Many industries are faced by disruption these years and the winners of the new market conditions seem to be the owners of the products and the owners of data. All in between, as retailers and distributors are subject to large changes. Their value add to the products is limited and is easily replaced by new digital trade platforms. The new online trade and the decreasing market shares of the retailers is also a challenge to the distributors and the logistics businesses.

Companies' Efforts in the Field of Transport Should Be:
- Active in exploiting the new digital meeting forms created during corona and prioritizing when physical meetings are needed across borders and across countries.
- Develop guidelines for employees towards the use of more public and shared transport when this applicable, and with the planned expansion of the rail network in Europe, a large part of passenger transport can be moved from planes to trains with approximately the same travel time. Night trains will again becoming an attractive mean of travel.
- Quickly convert companies' fleets to electric cars—it is not only sustainable and responsible, but also cheaper.

Climate Nexus for Sustainable Transport

Transport and logistic must change to protect climate change, but also to solve some of the other challenges that transportation of people and goods are causing. The evolution of transportation has been speeded up by globalization and we need to reinvent efficient means of transport and reintroduce energy efficiency here, by:

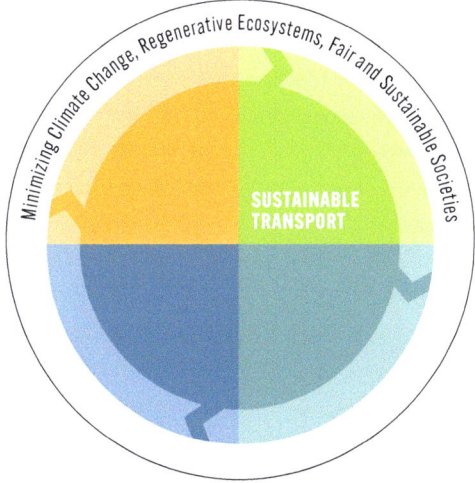

Redesign of collective, public, electrified transport in cities contributing to smart, livable, and sustainable cities.

Minimize particle and noise pollution that causes healthcare issues for millions of people.

Solve congestion of traffic in cities to make space for increasing urbanization and make cities livable with recreative areas, local food production and biodiversity.

Reference

IRC/PBL. (2013) (2010). *Direct GHG emissions from transport sector (annex II.8)*. IEA.

Chapter 11
Transition to Sustainable Land Use, Agriculture and Healthy Diets

Land use is a large contributer of GHG-emissions but in addition to impacting the climate, agriculture and deforestation also create challenges, as:

- Food waste and overconsumption of food.
- Clearing wild nature, especially for meat production, challenging wild nature and biodiversity to the detriment of humans and the ecosystems we depend upon.
- Industrialization of food production creates unhealthy diets.

The most violent GHG emissions are those from the agricultural processes and deforestation emitting methane and nitrous gasses. They have increased dramatically since the 1950s when we were just 2.5 billion people on this planet, all with a much more plant-based diet. Food production, especially meat production and food waste has increased tremendously over the last decades, which causes an additional increase in GHG emissions. Much more land has been put under cultivation, and deforestation has increased together with an increase in the use of pesticides. Deforestation is caused by development of new agricultural land, grazing, mining, and drilling. Especially tropical deforestation is now causing GHG emissions on the same level as the total CO_2 emission from the EU, Russia, and Japan—so it is severe. However, it is important to remember that deforestation has been a precondition for the spread of man over hundreds of years.

Since the 1960s population growth rate has increased more than deforestation. This is due to the green revolution and the introduction of fertilizers and pesticides, especially in the western world and Asia. In some regions the green revolution four-folded the crop yields and lowered the need for deforestation. Then crop breeding programs were intensified also as a part of the green revolution resulting in a strong spread of monocultures without weeds. This gave a very efficient agriculture but resulted in a rapid decline in biodiversity and kicked off the biodiversity crises we are facing now. This highly efficient agriculture based on monocultures results in disappearance of wild plants and animals throughout the food chain.

© The Author(s), under exclusive license to Springer Nature Switzerland AG 2024
G. Haar, *The Great Transition to a Green and Circular Economy*,
https://doi.org/10.1007/978-3-031-49658-5_11

Deforestation has throughout human settlement and expansion been the source to land, food, and natural resources. Forest has always been the first victim in the spread of humans. Deforestation throughout history have had a tremendous climate impact for centuries. Forest holds huge amount of carbon in their biosphere both above and below ground. The below ground part has historically been underestimated even though it was the basis for fertile soils because so much debris of organic matter was released by deforestation. See more on climate impact from deforestation in Chap. 3. Wild forests and jungles are a basis for high biodiversity and important in both creating carbon storage and biodiversity.

UN agreed upon a new biodiversity strategy to strengthen climate action (UNFCCC.INT) at Biodiversity COP 15 held in December 2022. Here Governments committed to protect 30% of land and water considered important for biodiversity by 2030. Currently, only 17% of terrestrial and 10% of marine areas are protected. Nations, large associations, and corporations has committed to the pledge from BD-COP15. The goals set for the EU on transition of agriculture and creating regenerative nature for biodiversity are seen in Fig. 11.1 to meet the UN commitments.

EU Nature Restoration Plan has made following key commitments by 2030:

1. Legally binding EU nature restoration targets to be proposed in 2021, subject to an impact assessment. By 2030, significant areas of degraded and carbon-rich ecosystems are restored; habitats and species show no deterioration in conservation trends and status; and at least 30% reach favorable conservation status or at least show a positive trend.
2. The decline in pollinators is reversed.
3. The risk and use of chemical pesticides is reduced by 50% and the use of more hazardous pesticides is reduced by 50%.
4. At least 10% of agricultural area is under high-diversity landscape features.
5. At least 25% of agricultural land is under organic farming management, and the uptake of agro-ecological practices is significantly increased.
6. Three billion new trees are planted in the EU, in full respect of ecological principles.
7. Significant progress has been made in the remediation of contaminated soil sites.
8. At least 25,000 km of free-flowing rivers are restored.
9. There is a 50% reduction in the number of Red List species threatened by invasive alien species.
10. The losses of nutrients from fertilizers are reduced by 50%, resulting in the reduction of the use of fertilizers by at least 20%.
11. Cities with at least 20,000 inhabitants have an ambitious Urban Greening Plan.
12. No chemical pesticides are used in sensitive areas such as EU urban green areas.
13. The negative impacts on sensitive species and habitats, including on the seabed through fishing and extraction activities, are substantially reduced to achieve good environmental status.
14. The by-catch of species is eliminated or reduced to a level that allows species recovery and conservation. https://eur-lex.europa.eu/legal-content/EN/TXT/?uri=celex%3A52020DC0380

Increasing organic farming and biodiversity-rich landscape features on agricultural land	Halting and reversing the decline of pollinators	Restoring at least 25 000 km of EU rivers to a free-flowing state	Reducing the use and risk of pesticides by 50% by 2030	Planting 3 billion trees by 2030

- **30% protected land and oceans in EU**
- **10% wild nature – strictly protected**
- **20 billion €/year for regenerative biosystems**
- **Leading position within biodiversity based on UN's new global framework.**

Fig. 11.1 EU Biodiversity strategy. 30% protected land and oceans in EU. 10% wild nature—strictly protected. 20 billion €/year for regenerative biosystems. Leading position within biodiversity based on UN's new global framework

The objective of the EU Biodiversity strategy (EU Comission, 2020) is to build societies resilience to future threats such as climate change, forest fires, food insecurity, and disease outbreaks. This include protecting wildlife and fighting illegal wildlife trade. The Natura 2000 areas will be enlarged, and the monitoring will increase. EUC has proposed the first EU restoration Law and has made the very bold commitment at the UN COP15—biodiversity conference that the EU will become the most biodiverse region on the planet, which is very ambitious coming from being the least biodiverse region with decrease both on land and in oceans. EU has committed to the pledge for each country to avoid the averaging out of poor nature and biodiversity against high biodiversity areas in Europe. This challenges some of the countries with extensive agriculture as Denmark, Finland Netherlands, and Belgium.

This interacts with the EU Farm-to-Fork strategy on transforming the European agriculture and farmers to become active manager of nature as well as sustainable food producers. Figure 11.2 illustrates the means of the Farm-to-Fork strategy that aims to accelerate the transition to a sustainable food system that should (EU Comission, 2022):

- have a neutral or positive environmental impact.
- help to mitigate climate change and adapt to its impacts.
- reverse the loss of biodiversity.
- ensure food security, nutrition, and public health, making sure that everyone has access to sufficient, safe, nutritious, sustainable food.
- preserve affordability of food while generating fairer economic returns, fostering competitiveness of the EU supply sector, and promoting fair trade.

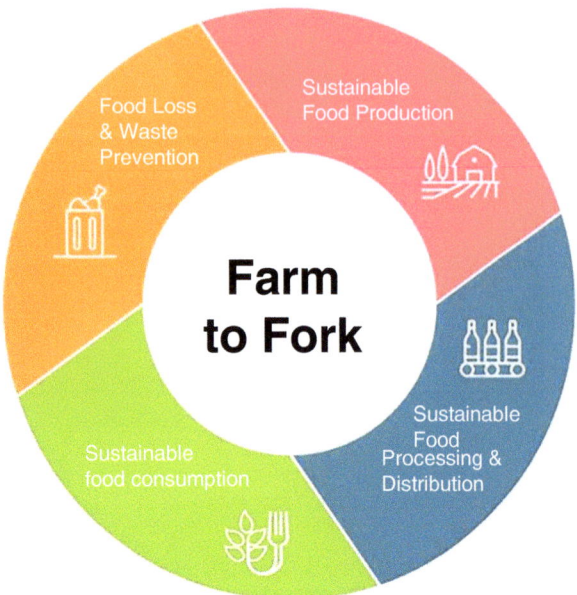

Fig. 11.2 **EU Farm to Fork.** https://food.ec.europa.eu/horizontal-topics/farm-fork-strategy_en

All this will also result in a new circular bioeconomy to restrain the values from bio-resources and utilize more components from the biomass and the organic materials, as proteins, fibers, lipids, before composting for energy and nutrient recovery. The industry and the landscape in Europe are facing tremendous change over the years to come.

In many countries over the globe, as well as in Denmark, a very large part of the land is agricultural land. In Denmark agriculture accounts for over 60%, with the increasing spread of large-scale farming or industrialized agriculture, as seen many places in Europe, North America, and Asia. The idyllic smallholder farms no longer exist or only to a very limited extent as hobby farms in the industrialized part of the world. The industrialization of agriculture over the last 50 years has abandoned the wild nature and introduced large and expensive machines that can be operated by few. Efficient agriculture has destroyed wild fences, wild grassland, streams, small forests, ponds, and wetlands where there used to be plenty of wild nature. This was the kickoff of the biodiversity crisis we are now facing and that threaten our own existence.

The reasons for loss of biodiversity are destruction of habitats, pollution, overfishing, overhunting, but also climate change cause loss of biodiversity of plants, animals, fungi, algae, and microorganisms. The loss of ecosystems is an even larger threat to biodiversity because regeneration becomes more difficult when systems as the coral reefs are totally disappearing. Biodiversity can also be harmed by the introduction of non-native species to other habitats because that can tilt the balance of an ecosystem. Read more about biodiversity in National Geographics education articles: https://education.nationalgeographic.org/resource/global-biodiversity/.

The Living Planet Index (LPI)—which tracks populations of mammals, birds, fish, reptiles, and amphibians—reveals an average 69% decrease in monitored wildlife populations since 1970. The 2022 LPI, by WWF analyzed almost 32,000 species populations. It provides the most comprehensive measure of how they are responding to pressures in their environment. Biodiversity loss by region (LPI) are:

- Europe and Central Asia 18%
- North America 20%
- Asia 55%
- Africa 66%
- Latin America and Caribbean 94%
- Source: WWF Living Planet Report, 2022. Building a nature-positive society (https://livingplanet.panda.org/)

The reason why especially Europe is so low in loss is because the biodiversity loss here started before the measure, that started in 1970. Population growth, deforestation, industrialized agriculture was developed here before the 1970ies, and the loss of biodiversity began here much earlier than other parts of the world. North America still has low population density and holds enormous amounts of wild nature counting in Alaska and Canada. The Asian population has developed since 1960ies as seen from Fig. 11.3 (Hyde, Gapminder, UN, 2017/2023/2022). Asians still live less urban and industrialized, and still with a large traditional rural population. The industrialized agriculture is not so widespread here as in Europe and North America. Latin America and Caribbean are really experiencing the results of deforestation since the 1970ies mainly due to the spread of protein production and palm oil production as input for meat production and industrialized food. But also due to logging from the forests. The population size in South America is not the challenge here, the transferal of forest into agricultural land by multinational companies providing inputs for other regions that South America.

Monitored freshwater populations have seen an alarming decline of 83% since 1970, more than any other species groups. Habitat loss and barriers to migration routes account for around half the threats to these populations.
   — WWF, 2022.

Meat production has increased tremendously over the last decades and is causing not only GHG-emissions but loss of biodiversity and wild nature. Figure 11.4 shows climate impact from various food products in the full value chain.

Food is now consumed on a scale that obesity has become one of the major global health problems. The combination of cheaply processed carbohydrates and animal fats seems to be one of the large culprits. The human body and brain need fats and amino acids, which many today get from meat. Some kinds of meat also contain a lot of unhealthy fatty acids and fats, compared to the plant-based fatty acids or fatty acids from fish, which is probably the diet we are adapted to by evolution.

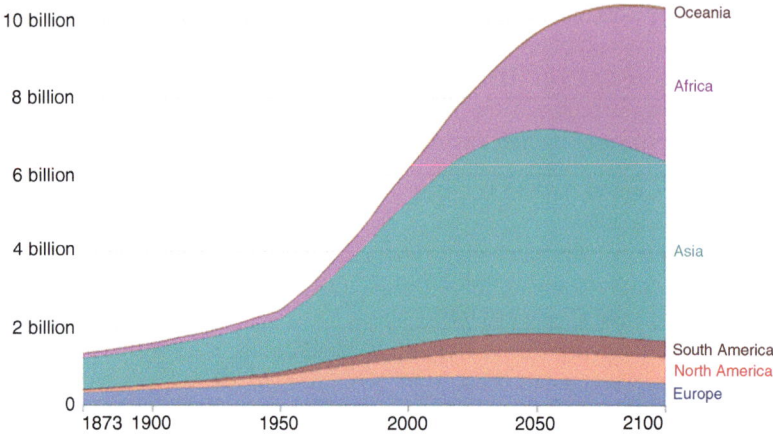

Fig. 11.3 Population by Region (1873–2100). Historic estimates from 1950 to 2021 and projected to 2100 based on *UN medium-fertility scenario.* Note: Historic country data are based on today's geographical boarders. *Our World in Data. HYDE (2017) Gapminder (2023) UN (2022).* https://ourworldindata.org/grapher/population-regions-with-projections?time=1873..latest

There are some interesting correlations between high consumption of food, especially meat and easy processable carbohydrates, and some severe lifestyle diseases such as diabetes and bowel cancer. It is important to note that 800 million people suffer from obesity, causing more deaths than hunger does today.

Hunger is still a severe problem and will increase with climate change. Hunger can rarely be put down to lack of food production. Hunger is caused by lack of distribution of food to where it is needed. This typically happens because of human conflicts and geopolitical challenges. Hunger is a human-caused situation that is not correlated to regional food productivity. Though severe droughts are expected in most regions and will cause lack of food production and famine, as a direct impact of climate change. Overproduction of food in some parts of the world has indirectly contributed to famine and climate refugees elsewhere. It requires sustainable and holistic systems to solve this and not more industrial food production.

Originally, fish and shellfish were the most common protein source for humans. These proteins and fatty acids were the basis for evolution of the brain of homo sapiens. Other types of meat were only slowly introduced to our diets later in the history of human evolution (1000–2000 years), probably not with any significant impact on our biological evolution. Humans have not become predators; we are still omnivorous with focus on the plant-based diet.

Figure 11.5 shows the impacts from agriculture on the environment measured as greenhouse gas emissions (GHG), land use, freshwater consumption, nutrient emissions (eutrophication) and biodiversity. The production of farmed animals, mainly the ruminant (cows and sheep), has a large share of impact on the climate from agriculture, and there is a big difference in the footprint of the different animal species.

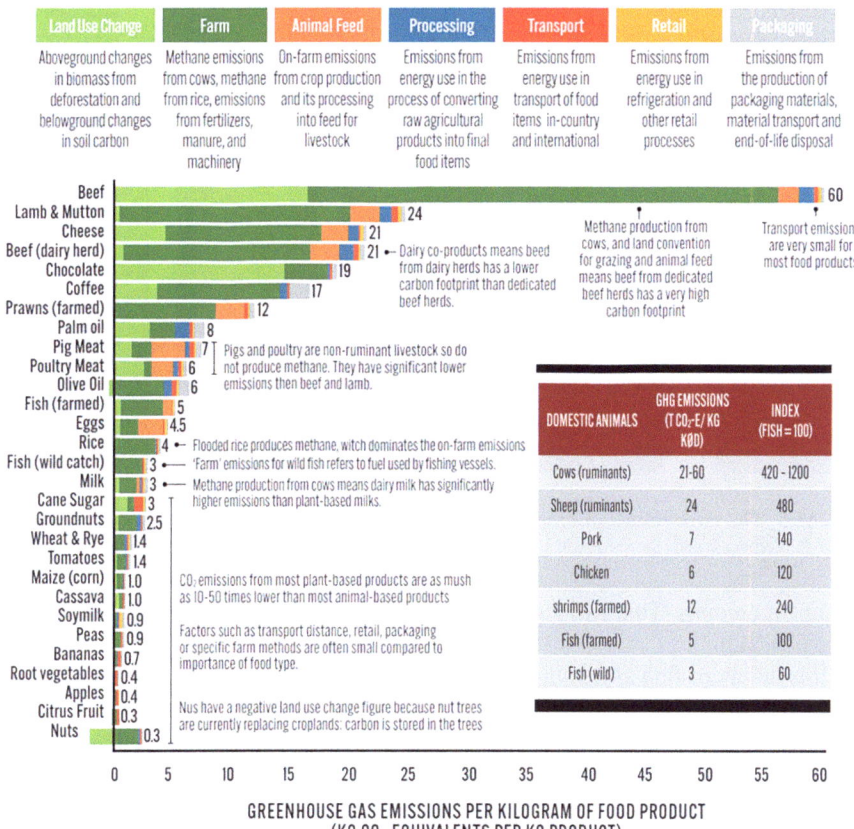

Fig. 11.4 GHG emission in the value chain from Food Products. The increased meat production is not only due to an increasing population, but the average meat and food consumption in the rich part of the world has also increased enormously. In Denmark, we have a high yearly consumption of meat, i.e., 52 kg/citizen. The US has the highest with 100 kg/citizen and India the lowest consumption with 5 kg/citizen

From a historical perspective the past 1–5000 years, meat has been of great importance and for the last 50–100 years it has had a tremendous significance. Although 1000 years is no time in an evolutionary perspective, the meat production has obviously had a tremendous impact on the development of the Anthropocene age, because it greatly accelerates human dominance of the land areas. It takes a huge amount of land to feed all the livestock that humans eat, and it has a large impact on nature, on GHG-emissions and on biodiversity. Figure 11.5 shows how agriculture affects its environment in terms of GHG emissions, land use, freshwater consumption, nutrient discharge (eutrophication) and biodiversity.

The impact from every person is large affected by what we put on our plates. We do not all need to become vegans or vegetarians for our diet to become more climate friendly. We should reduce meat consumption—especially meat coming from

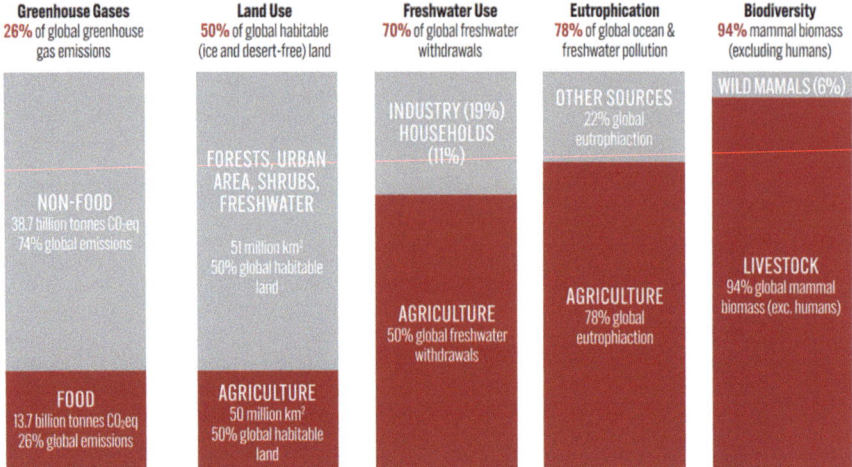

Fig. 11.5 The environmental impacts from food production and agriculture. Data sources: Poor & Nemecek (2018); IN FAO; UN AQUASTAT: Bar-On et al. (2018). Ourworldindata.org

ruminant animals. This will significantly reduce the climate impacts and improve human health.

It is an interesting aspect that the Japanese diet to a high degree consists of fish and vegetables, and many Japanese still live on an old life philosophy of not eating animals with four legs. The Japanese diet may be one of the reasons that they experience an average life expectancy of 84 years, being one of the highest in the world. In many western, industrialized countries the life expectancy is still below 80 years, despite the well-developed health care systems and high health costs these countries hold. The European Mediterranean countries and Canada also appear to have found a formula and a way of life that provides longevity. Around the world are some small communities called Blue Zones where people tend to live longer, healthier and turn 100 years at a much higher rate than elsewhere. Research are being conducted on what are the common denominators in these Blue Zones, as preferably plant-based diets, daily exercise, and important social relations.

Studies show that prefabricated food is less healthy than locally produced food and the EU has now (2021) regulated the industrially produced food, in relation to the use of trans-fatty acids (Comission, 2022; WWF, 2022; Poor & Nemecek (2018); Bar-On et al. (2018); Commission, Transfedtsyrer i fødevarer, COM(2015)). When people cook their own food from locally produced inputs, they are healthier almost regardless of their diet. In that perspective still more than 50% of the food consumed globally is locally produced—so industrialization (of food production) is rather a part of the problem than for the good.

When developing countries have a low life expectancy, it is probably due to factors outdoing the diet such as lack of access to clean drinking water, lack of healthcare and hospitals, and high child mortality due to infectious diseases and malnutrition as well as many traffic accidents, which is still a big killer here. It also becomes apparent that we have reached a point where more people die of obesity than of malnutrition, which is a sad statistic. Unfortunately, there is a correlation between poor health, lifestyle diseases and access to too much food. Telling a story of when humans gain wealth, we create some peculiar problems for ourselves and our management of nature.

Meat production has had an enormous impact on the development of the earth's geology and the Anthropocene Age, because animal husbandry has contributed greatly to human dominance. It takes an enormous amount of land to feed all the livestock that people eat. Agriculture accounts for half of the accessible land on the surface of the planet—out of this animal production accounts for 77% of farmed land. That has a profound effect on natural biotopes and therefore agriculture causes the most significant negative effect on biodiversity. See Fig. 3.4 earlier in this book.

The industrialized meat production requires a large amounts of plant proteins (soya) as feed, typically imported from regions outside EU. Grasslands and the transformation of original forest to agriculture for animal feed proteins together with the production of palm oil are currently causes the largest decline in biodiversity and GHG emissions.

Monocultures in agriculture both of plant varieties and domestic breeds increase the risks of vermin, diseases, based on viruses, insects, and bacteria. Therefore, wild nature is needed to create resilience and genetic variation for future food production and breeding of varieties. Monocultures, lack of biodiversity and uniformity, combined with overpopulation and our close dealings with livestock and wildlife, increase the risk of pandemics such as the coronavirus. As a result, people's way of life will only become more and more vulnerable, and our dependence on wild nature will increase with population growth. Saving the biodiversity on the planet will require a drastic change in our diets in parts of the industrialized world.

Another large challenge from our food production is the amount of food waste. Very long value chains of food production create food waste and overproduction estimated to up to one third of all food produced. Example of long value chains here are fruits and vegetables that are grown and harvested on one side of the planet and consumed on the other side of the planet with very many distributors and retailers before the fruit end up with the consumer. The same goes for fish, where 5–6 distributers touch the fish between the fisherman and the consumer. This makes the food canker and result in large amounts of food waste.

The share of food waste must be reduced significantly—especially the industrialized food value chains create a large portion of food waste when the goods reach the consumer. Low prices on food encourages to careless purchase and handling of food.

Nature and natural resources are owned by all of us and should also be managed with respect for nature. The future generations depend on nature and natural resources, and therefore there is a strong need for a thoughtful and deep transition of our agriculture and our food habits.

Companies also have a major task in contributing to the transition to sustainable agriculture with a wild and biodiverse nature, by:
- Transition to a circular bioeconomy, where all resources are used sustainably and not depleted. When natural resources have been harvested, the most value must be extracted from them, and here food is the most value-creating.
- Offer sustainable food products where the impacts throughout the value chain can be accounted for.
- Reduce waste throughout the value chain, and this applies to both food waste and packaging waste, but also waste and downgrading of products that could have been better utilized.
- Create proximity in food production and closer interaction between producers (primary producers) and consumers to promote the demand for sustainable food production when we see and understand how animals and food are produced and handled. The absence of pesticides is an important part of this transition.
- Ensure that no harmful chemicals or pesticides are added to the products or packaging so that we achieve clean material loops from farm to fork and for bio composting to return nutrients to the soil.
- Choose and offer a balanced, healthy diet that does not affect the climate and nature unnecessarily. Often the climate-friendly solution will also be the healthy solution.
- Support the development of wild nature and biodiversity to ensure a sustainable world for our fellow humans.

Climate Nexus on Sustainable Agriculture, Land use and Healthy Diets

Creating a sustainable transition for for agriculture, land use and healthy diet

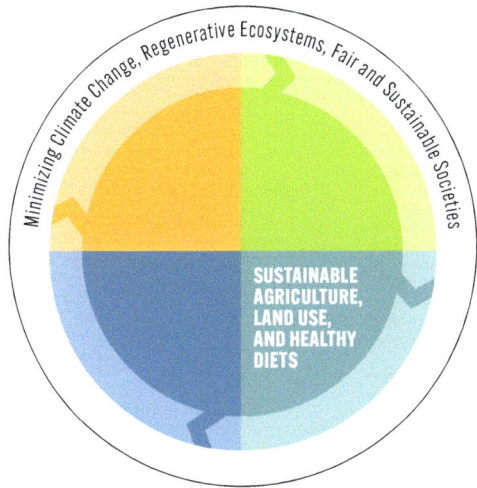

Re-establishment of natural habitats to rebuild biodiversity globally.

More balanced diets lead to healthier and more affordable food, minimizing obesity and other diseases that strain humans and our economy today.

A closer connection between food producers and food consumers will accelerate the demands for more sustainable and local production since we will again know how animals are treated and how pesticides are added.

Urban Farming to green the cities, make food production come closer to people and create an understanding of methods and influence on biodiversity. Taking the cars out of the Cities will leave space for Urban Farming since we will need much less space for transport.

Intelligent use of bi-products and organic waste in a cascade before bio-composted. There are huge potentials in extraction proteins, fibers, and lipids from organic waste and there is a huge potential in using organic bi-products and waste for feed into the production of new types of meets and proteins from worms, insects, fungi, etc.

References

Data for Population by World Regions based on HYDE (2017), Gapminder (2023), UN (2022). https://ourworldindata.org/grapher/population-regions-with-projections?time=1873..latest

EU Commission. Trans fatty acids in foods. COM(2015)619, (2015). https://www.foedevarestyrelsen.dk/Leksikon/Sider/Transfedtsyrer.aspx

EU Comission. (2020). *EU Biodiversity strategy for 2030*. Retrieved from Environment, European Comission: https://environment.ec.europa.eu/strategy/biodiversity-strategy-2030_en

Hyde, G., & UN. (2017 / 2023 /2022). *Our World in Data*. Retrieved from Population by World Region: https://ourworldindata.org/grapher/population-regions-with-projections?time=1873..latest

Poor & Nemecek. (2018). Reducing food environmental impacts through producers and consumers. Globale data given from 38,700 viable farms in 119 countries. Science. Image source from Noun Project. Our World in Data. https://ourworldindata.org/food-choice-vs-eating-local

Poore & Nemecek & Bar-on et al. (2018). The environmental impacts calculated as greenhouse gas emissions, land use, water consumption, nutrient emissions. UN FAO + UN AQUASTAT. https://ourworldindata.org/environmental-impacts-of-food

WWF. Living Planet Report. (2022). https://livingplanet.panda.org/en-US/

Chapter 12
Driving the Climate Nexus: People and Money

To transform into a Green and Circular Economy requires a good understanding of the Climate Nexus and the solutions presented in this part II and illustrated in Fig. 12.1. This Climate Nexus is an overview of all the solution presented earlier in this part II of the book. Implementing all this needs people and money as the drivers of the transition. The drivers on a society level are:

- Cooperation between all actors in society and along value chains, with the consumers and businesses as the key transformers.
- Dissemination of knowledge and building of competence on the Green and Circular Economy in details for all the steps in the transformation to happen.
- Financial instruments to price the use of externalities.
- Capital for more risky investments in new business models, new technology and new markets.

The transformation is as large and extensive as the digital transformation and applies to companies, young and old, politicians, municipalities and so on.

There is no quick solution from some central position. It is a long haul that we need to take together over the next 10 years. Strongly supported by the EU's Green Deal and the Roadmaps here. This chapter reviews the actors in the transition that will help create this new society and the new economy that drives new infrastructures and new business models at all levels. The need for capital for the green transition is also discussed here, and in part II of the book is a method that companies can use when transforming their business model to Circular Economy.

Transition to a Green and Circular Economy will involve all levels and all actors in the society and it requires a structural change of legislation and infrastructure. This is a complex task and challenges the linear industry thinking and the silo-based structures of society, but especially challenges the roles of public authorities, because they to a much greater extent must become sparring partners and involve companies and experts to find the solutions of the future instead of just being

© The Author(s), under exclusive license to Springer Nature Switzerland AG 2024
G. Haar, *The Great Transition to a Green and Circular Economy*,
https://doi.org/10.1007/978-3-031-49658-5_12

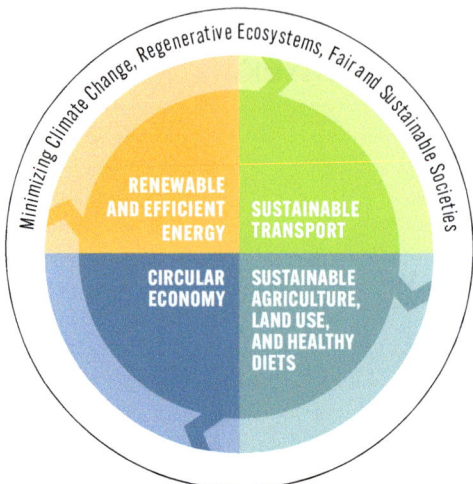

Democratic and free access to energy for all resulting in political stability, spread of democracy, and minimizing poverty.

Predictable prices on energy supply as an important resource for consumers and business.

Eliminating pollution from combustion engines That causes sever health care problems, from particles and noise.

Restoring scarce resources by redesigning, reusing, and recycling.

Counteracting environmental disasters seen globally due to tremendous amounts of waste being exported from the consuming countries and deposited in the developing countries, resulting in deposits of waste as plastic, textiles, electronics with the result of pollution and other environmental disasters.

Replacing products and quality designed for the linear economy with products that are designed for the circular economy, thus ncreasing the potential for a sustainable lifestyle of future generations.

Controlling and replacing chemical pollutants from material streams and pollutants introduced earlier without accounting for the healthcare and environmental problems they caused.

Creating immediate energy savings from recycling compared to extraction of virgin resources.

Creating regional value chains that are transparent and traceable to ensure sustainable consumption to stabilize and develop regional economies and wealth.

The basis for regenerative ecosystems due to the decoupling of land use for mining of virgin resources.

Redesigning the global value chains to regain control of economy, and security of supply.

Redesign of collective, public, electrified transport in cities contributing to smart, livable, and sustainable cities.

Minimize particle and noise pollution that causes healthcare issues for millions of people.

Solve congestion of traffic in cities to make space for increasing urbanization and make cities livable with recreative areas, local food production and biodiversity.

Re-establishment of natural habitats to rebuild biodiversity globally. More balanced diets lead to healthier and more affordable food, minimizing obesity and other diseases that strain humans and our economy today.

A closer connection between food producers and food consumers will accelerate the demands for more sustainable and local production since we will again know how animals are treated and how pesticides are added.

Urban Farming to green the cities, make food production come closer to people and create an understanding of methods and influence on biodiversity. Taking the cars out of the

Cities will leave space for Urban Farming since we will need much less space for transport.

Intelligent use of bi-products and organic waste in a cascade before bio-composted.

There are huge potentials in extraction proteins, fibers, and lipids from organic waste and there is a huge potential in using organic bi-products and waste for feed into the production of new types of meets and proteins from worms, insects, fungi, etc.

Fig. 12.1 Climate Nexus. An overview of the elements described in part II of this book is summed in this Climate Nexus illustration

> The green transition involves all of society in new ways. It calls for break-up of old structures and innovation all over.

controllers. The authorities also have a responsibility to rethink broad solutions for all in society, and not just for the large corporations that are good at lobbying.

Figure 12.2 illustrates the stakeholders and their complex interaction, and the different stakeholders are described here (Haar 2021).

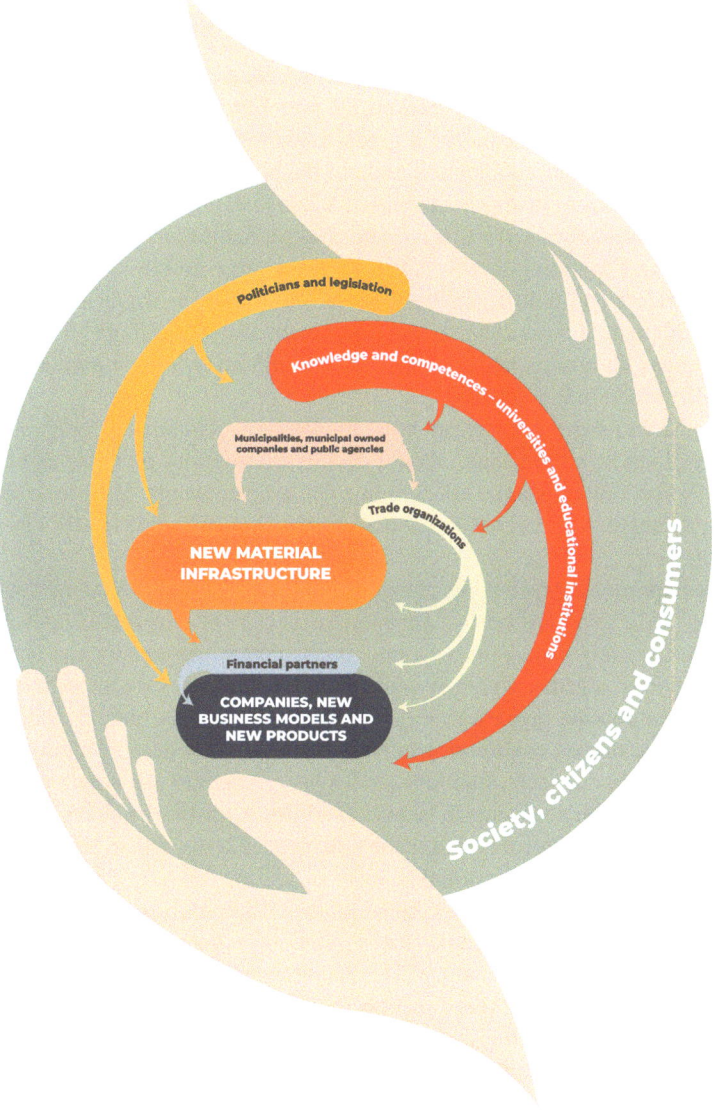

Fig. 12.2 Stakeholders in a green and circular economy

The citizens and consumers are the central stakeholders in this transition because it is about our welfare, our safety, and our future. When consumers demand sustainable products, they greatly influence companies to a more sustainable, traceable, and transparent production. Of course, there are also consumers for whom a cheap price is the most important purchasing parameter, but more and more people are engaged in the green transition. Just as more and more people reject the take-make-waste business model in the linear economy.

Citizens should first and foremost change their own way of consuming, travelling, and requesting transparency and traceability from producers, and they should be willing to pay for long-life products with high quality, and discard single-use products. They can and may do this already tomorrow.

The citizens are the voters, consumers, inhabitants in cities and areas where change is needed. The citizens have great influence on the politicians and each other. The elections and demonstrations all over Europe and the rest of the world clearly show that movements for a green transition and a sustainable planet are strong. People are worried about the state of the planet and the climate, and people want the politicians to take responsibility for the challenges we are facing.

International studies show that climate change, natural disasters, environmental challenges, and declining biodiversity are what worry people in general, but also business leaders and politicians the most. The importance of green transition surpasses challenges such as cybercrime, rising inequality, and refugee flows.

Consumers should implement a Consumers Hierarchy before purchasing products as illustrated in Fig. 12.3.

The model in Fig. 12.3 correspond to the Resource Hierarchy in Fig. 9.1 earlier in this book on how companies should decide on design of products and business models in becoming resource efficient in a Circular Economy. If these to pyramids are implemented with consumers and in companies, this will totally change our economy.

The companies are facing big change, primarily because the customers want sustainability and green transition. Companies are meet by new legislation and new market conditions on demands for recycling, transparency, sustainably and responsibly produced goods, together with new demands from employees for meaningful jobs.

Investments in the green transition are highly dependent on companies having safe and stable market conditions and knowing future changes and regulation. Eco-labelling, taxation, tender conditions, and subsidy schemes are important for companies to know of in good time to be able to adjust. Adjusting products and business models is not the large problem. It is the time needed to be able to adjust to the new market conditions that is important.

Political hesitancy in the green transition in some countries, as in Denmark, creates uncertainty among companies and inhibits investments and the readiness for change in the business environment. The countries where the political level is weak in the framing and setting the direction towards a Green and Circular Economy together with business will lose position globally. Contrary to popular belief, many of the solutions lie within the companies, and sustainable companies are not afraid of taxes or environmental regulation if it is part of an overall transition and creates new market conditions for all.

Fig. 12.3 Consumer patterns hierarchy

In EU, politicians are willing to legislate and set a strong roadmap for the great transition to a Green and Circular Economy. The countries that are agile and where the political level can work together with companies on the transition, this creates a first-mover position for the companies on the EU inner green market and globally. Countries that are seen progressive in this transition are typically the old industrial countries as France, Germany, Holland, etc. Also, some states in the US are progressive in the transition, also approaching European companies and experts to work on the transition. The Nordic Council has announced that the Nordic Region will become the most sustainable region in the world, but when it comes to Circular Economy and sustainable consumption there is still far to go. The consumption and waste production here is of the largest on a global measure.

Trade associations and agencies play an important and facilitating role, as they must connect to legislators, municipalities, and companies to create the new circular infrastructure for the clean material loops. They are the important connection between municipalities and public enterprises, legislators, and companies in designing the new systems and business models on the material level.

The trade associations are important in facilitating knowledge on new legislation and EU roadmaps to the companies. They are also policy makers and should contribute to designing a legislation that support the transition. In that way the financial incentives to recycle, to invest in renewable energy, and to reduce climate and environmental impacts from our consumption are crated. Trade associations need to improve in connecting members with knowledge institutions and breaking down some of the silos that are built up in the linear economy. In some cases, this is successful, but often there are a lot of parallel initiatives trying to solve the same challenges within their own silos.

The ***New Material Infrastructure and New Material Loops*** are the result of the strong interdisciplinary and cross-sectoral collaboration that is very much needed. Read more about this in Chap. 9.

In many countries the ***municipalities and public enterprises*** handle our household waste, and the waste from SME's. Their position, their infrastructure and business models also need to change. This should be done in close cooperation, as in Private Public Partnerships (PPP) between municipalities, public enterprises, and the recycling industries. The recycling industry is often subcontractors to the municipality and the public-owned enterprises and collect, handle, and sort our waste. In the transition from waste handling towards material loops and material banks especially this industry is an essential player in creating new ways of handling waste or since resources. In many countries waste legislation prevents the different stakeholders from changing from handling waste to handling resources. Here the new EU regulation is important to set the bar for the national changes needed to translate the recycling targets into action especially in the municipalities and public enterprises.

The municipalities and the public-owned enterprises also determine the infrastructure in our towns, and the cities and towns are a central platform for the transition. It is here the citizens live and move about, where waste is being handled, and products consumed. At the same time, it is in the towns and cities that the effects of climate change as rising waters, changing weather, and drought are challenging the citizens. Massive climate adaptation must happen to preserve the cities and towns as sustainable and green residence for the citizens, and to preserve the value of the citizens' homes and buildings.

Involvement of citizen locally by the municipalities is central to find the future solutions, and the level of involvement is different. First and foremost, it requires design of new infrastructure for the materials and material loops, and then standardization in the collection, handling, sorting, and quality assurance of the materials is crucial. Here, the public agencies play an important role, being the link between legislation and implementation of the green transition. The state as well as the municipalities must contribute by enabling the public purchasing power as a driver for the green transition. Today, there is a large difference across countries on how the public sector embraces the transition at all levels. How the manage legislation and the intentions towards the Green and Circular transition. Some countries are in the forefront of setting new and innovative requirements in their tenders and how they manage the city planning laws.

In countries where municipalities depend on household waste as a supply for local incinerators to generate district heating, as in Denmark, there is a contradiction between the need for a Circular Economy and the local heating economy. This contradiction of interests will slow down the transition to a green economy.

Knowledge- and educational institutions, such as universities, vocational schools, and Approved Technological Service -institutions, need to work closer with the companies to transfer knowledge and create the innovation necessary. The green and circular transition is a knowledge-intensive and technical transition, and especially the knowledge institutions' cooperation with the SMEs is often a challenge for both parties.

New technology, the understanding of materials, and documentation of the impacts from companies and products must be transferred from knowledge institutions to the companies. It works for the large companies that employ the academic staff with the technical understanding as responsible for the transition. Academic skilled people in the companies can collaborate with the knowledge institutions and universities and good at defining their needs. However, achieving the same synergy between knowledge institutions and SMEs is often a challenge, and the need here is large. The gap is larger than ever before if we are to succeed with the green transition. Unfortunately, it seems that knowledge institutions have thrown themselves at entrepreneurs – both in their own ranks, among researchers, but also broadly.

Entrepreneurship centres (garages) open at all universities, but ongoing contact between researchers and the existing companies, and SMEs, is often limited. Thus, we miss out on the transformation of a large part of the existing companies, despite their innovative power and adaptability to new market conditions. It is a challenge when too little attention is given to the existing SMEs and the need for developing competence both with SMEs and researchers and a potential for accelerating the transition is overlooked in the recognition that 70-80% of the business environment consist of SMEs in EU.

The politicians must change the *legislation*, so that the new goals can be set for a Green and Circular Economy, especially that the material loops of recycled materials must take place on commercial terms, but also ensuring private investment in RE, not to mention creating regenerative ecosystems. The implementation of Circular Economy varies a lot across EU and a lot of change is driven by European politicians with the new Green Deal and comprehensive legislation on Circular Economy and the new EU taxonomy on sustainability, defined as ESG – Environment, Social and Governance. In this way, the national politicians must introduce national legislation for the implementation of the ambitious EU-legislation on the green transition.

There is a need for national legislation on minimum sustainability requirements, procurement guidelines for public procurement and taxation on products not aligned with the new EU-taxonomy on sustainability. Political foresight on introducing legislation and taxation is important to create stable market conditions and the good business cases in companies. Thus, achieving the European goals on Circular Economy, Climate Neutrality, Climate adaptation, Biodiversity and Wild Nature to protect land, oceans, and freshwater systems. It requires clear, long-term guidelines and legislation to direct private investments into the green economy.

The green transition is complex, very complex, because it involves all layers and silos in society. It requires a larger number of professional skills and the ability to work together across disciplines. Interdisciplinarity has been a buzz word for many years, but with the green transition, it is more relevant than ever before. There is a lack of competences that can drive complex and interdisciplinary transformation with very diverse stakeholders. In particular, the technical, scientific, and commercial competences must interact in new ways.

Denmark and the Nordics are known to hold an advantage with flat management hierarchies and tradition for informal cooperation and knowledge sharing, but there is still a long way before the Nordics becomes a pioneers and front runners in the Green and Circular Economy. The focus is still on technology development in selected silos or industries, such as wind turbines, the pharmaceutical industry, and the industry of bioeconomy. There is not yet a broad understanding of the need for boosting skills and the cooperations across and along value chains required to transform society, infrastructures, and the large number of small and medium-sized industrial enterprises.

All stakeholders in the green transition need to become wiser as to *what* and *how*. Many do know *why*, as climate change, resource scarcity, pollution, and declining biodiversity has affected most people. There is a big need for in depth education and innovation within the transition to a green economy and Circular Economy. There is much commitment among citizens, and much committed power in writing and in discussions among intellectuals about Circular Economy and the Sustainable Development Goals (SDGs), but there is a lack of systematic initiatives to pull all stakeholders together and make them into play.

The future holds change, but it is surprising that the present does not offer more investments in the well-known technologies that are able to accelerate the transition. People are very committed and environmentally friendly on their own behalf. There still seems to be a lack of basic understanding of how the green transition is implemented in both private companies and public institutions.

> A Green and Circular Economy is the battle ground for nerds, and the technical and commercial competencies must be combined in companies and throughout society to transform into a green economy.

Financial Instruments to Facilitate the Transition to a Green and Circular Economy

Financial instruments are as important as the human capital in the transition. Money makes the world go round and before we have the economic incentives to change behaviour and care for the planetary boundaries the change will be sporadic. We need to pay for the externalities that are impacted by human activity and consumption. In other words, the products we buy carry an environment and a social footprint that we do not pay for. We only pay for the design, manufacturing, distribution and retailing of the products. Soon companies must account for these footprints in their annual reports and on their products. Therefore, the prices on food, products and services need to include the environmental impacts and the social impacts in the full value chain. The new regulations on reporting and the EPR put on products is the first step in this direction.

Various financial instruments exist to increase prices and preserve products, which are important to consider in developing sustainable business models:

- Regulation on long durability and maintenance of products (Extended Producer Responsibility)
- Regulation on transparent and traceable production and material use (SBI, PEF and Due Dilligence)
- Taxation of single-use and disposal of products
- Taxation on Green House Gas Emission from products in the full value chain
- Taxation on use of virgin materials in product.
- Transformation of the EU-financial support to farmers and forestry from a production support to a nature management fee measured on wild nature and biodiversity.
- Ocean fee put on the industries dependant on oceans and freshwater ecosystems to become access to this very vulnerable wild natural resource.
- Easy and fair payment for power and heat produced by RE by private entities.

Most of these financial instruments are already being adopted in the EU as part of the EU Green Deal (2020), or in the pipeline of future legislation. EU countries are also implementing climate taxes on heavy industries and agriculture. All economic research and recommendations from various Economic Councils show that environmental taxes create the incentives to promote a rapid transition.

Sadly, there is no financial payment for the exploitation of externalities, such as natural resources, biodiversity, climate impact, natural ecosystems, and so. We now pay with climate change, lack of nature and biodiversity. The time has come to end companies access to freely sell products without taking the responsibility of their value chains, the quality of their products, or the exploitation of nature.

At the beginning of the industrialization externalities were inexhaustible, and companies have for centuries operated with free and endless access to raw materials and nature. Companies didn't have to account for the indirect impacts of extraction of resources or disposal of waste. All causing damages that was expected to go away over time. Therefore, pricing of externalities is not built into the traditional economic models. We are in the rapid process of depleting our natural resources, raw materials, biodiversity, and the climate buffering capacity, and therefore we all need to pay for the use of externalities. Either directly as a price on the products or indirectly, by enabling consumers to opt out of products with a large draw on nature and virgin resources.

We must also expect a future climate tax on products. This is only fair since a significant part of the emissions the EU is responsible for in due to our overconsumption and insufficient recycling of the products. So, if the ambitious climate goals announced by the EU and the Nordic Council are to be taken seriously, we must expect taxes, or at least labelling of products with the climate footprints. As suggested with the new PEF by EU Green Deal, consumers need to know that the footprints of the recycled and recyclable products are much lighter compared to those of the single-use products.

EU wants to build an inner green market by creating transparent and traceable product information on the impacts of externalities. Regardless of the consumers' desire to act and buy sustainably, it has not been possible to figure out what is sustainable, or which impacts the products have on us and our surroundings. A competitive marketplace where customers are free to choose sustainable and responsible products requires transparency and traceability of the products and materials at a completely different level than we know. Only then the consumer can make sustainable choices. This requires regulation and ecolabelling. The value chains and the greenwashing done by corporations are too complex for consumers to be able to choose genuine sustainable products. Companies must now prepare for a new level of documentation on non-financial data to enlighten the issues throughout the value chain, and this will affect the development of business models and products.

The mentioned instruments and regulation will promote the genuine, sustainable, and responsible companies and expose those who use greenwashing. Lots of companies today use sustainability and green claims without accounting for the claims, and sometimes even against their knowledge. As seen in the oil industry, the automotive industry, the construction industry, and others. Greenwashing has become so widespread that the EU's new ESG taxonomy has a direct aim to counteract this. Companies must document their efforts and products on up to 16 environmental parameters in a life cycle analysis used for benchmarking (PEF).

Investments for the Green Transition

Even though the companies have the willingness to invest, the transition requires venture capital and new skills from the venture partners. It takes funds to transform to new business models, new products, new production facilities, and building new market channels. Therefore, farsighted private investors and banks are needed that understand the new circular business models and the green transition. Access to venture capital for companies is key to the green transition. Today it is a challenge, especially for the existing SMEs to attract the investments needed for the green transition. They finance this themselves, and some of the SMEs that have the innovation, and the solutions are too indebted, and then society miss a potential if we cannot assist those companies.

A large problem is that many banks are busy meeting implementing the new EU-legislation and tidying up the sins of the financial crisis in the 2000s, instead of looking ahead. Banks and Capital Funds are meet by new requirements from the Sustainable Investment Directive (SID) and the new ESG-taxonomy on sustainability and the hope is that it causes change and not only new type of number in their Annual Reports. Reporting does not save the planet—action does. The SID comes with the do-no-harm criteria, and these are comprehensive when a company must have a positive impact on at least one of the six environmental criteria and must do no harm to any of the others. Let's hope that this will completely change the focus of the investment landscape, as intended.

Facing inflation and increasing interest rates might slow down the phase of the green transition even though Circular Economy is one of the solutions of the economic insecurity. Our long and global value chains, the resource scarcity of raw materials, and the dependency on production and raw materials from regions at war or countries with increasing political instability and inhumanity is best mitigated by a Circular Economy where recycled materials are handled locally and made available in regional material loops. Automatization and robotisation enables the European companies to more local small-scale production with a greater flexibility instead of long and unstable supply chains with huge overproduction of product never reaching the consumers. So, a transition to Circular Economy is a transition to a more stable and local green economy that can counteract inflation and increasing interest rates. The war in Ukraine has taught the Europeans that dependency on Russian gas and oil is dangerous and causes not only human and political instability but also financial instability. Now we need to mitigate the risks of being dependant on raw materials not available regionally, and this is well done be recycling all the materials today called waste.

Financial institutions like banks, equity funds, and pension funds hold capital and look for new ways of creating return for owners and customers. Hopefully pension holders and shareholders will demand that capital be invested in a green economy and not in the unsustainable and fossil past. There are strong trends towards focusing on investments in the green transition, especially from pension funds that need to meet customers' requirements for a sustainable future. However, focus is still concentrated on entrepreneurship and investments in listed corporations communicating transition to SDGs and sustainability.

Seventy to eighty percent of European businesses consists of family-owned SMEs that need to prepare for the green transition, and this is where the large potential lies. Institutional investors and investment funds have started to communicate about the Sustainable Development Goals (SDGs) and a little bit on Circular Economy, but there is a lack of action, skills development, funds.

> Investments must become a locomotive for the green economy instead of the impediment it is today.

The new Directive on Sustainable Investments (SID) include shares and investments such as real estate and other large, asset investments made by financial institutions. Initially, it will apply to large, listed corporations, but will quickly drill down to all types of companies and affect bank loan and other capital allocations to companies. The new EU taxonomy is expected to quickly become the new standard for how companies communicate and report on sustainability, since it is also part of the new Corporate Social Reporting Directive (CSRD).

> **The EU Sustainability Standards on Environment are on six impact categories (ESRS):**
> - Protecting the Climate and Adaption to Changing Climates
> - Sustainable use and protection of water and marine systems
> - Circular Economy
> - Protecting humans and environment from pollution
> - Creating regenerative and natural ecosystems to protect wild nature and biodiversity.

A transformation to Circular Economy is a completely traditional business development process. The difference is the change in the external conditions than are the main drives for the new business model. It is not about changing the operating model in an existing market with existing products. Is about a new blue ocean with new market conditions, new requirements to the products and new material streams.

The Green and Circular Economy is a different setting. The rules of the game are changing, and the business models are fundamentally changing. Both from the outside and from the inside. Therefore, it is important that the transformation is anchored in the boardrooms and that the boards members give it as much attention as they do to auditing and risk management today. In the future, the larger corporations should have a Green or a Sustainability Committee just like they have an Audit Committee. Perhaps it needs even more consideration because the Green Economy together with Industry 4.0 will disrupt the company's business model rapidly. It will threaten the existence of the company if not adapted in time.

Reference

Haar, G. (2021). *Circular Economy in a Company Perspective (danish)*. Quare.

Part III
Methods and Tools for the Transition to a Circular and Green Economy

Part III provides methods and processes for businesses to transform into sustainable and circular businesses that meet the new EU legislation and drives the changes towards a fair and sustainable planet as so clearly stated in the SDGs. The methods and processes are developed from years of experiences working with corporations in their transition to a Green and Circular Economy, and all methods and processes are tested with companies.

Chapter 13
Introduction to Part III

Part III includes tools and methods for companies to drive change on minimizing impacts, and manage the targets and actions needed for the new market conditions in a Green and Circular Economy. It includes an introduction to companies on how to approach their work within sustainability and how to prioritize their efforts. It explains the UN Climate Protocol and its three scopes, it introduces a Materiality Assessment based on company impacts in the full value chain and gives hands-on tools and methods on how companies actively can participate in the great transition to at Green and Circular Economy and harvest all the potentials and wins that the new market conditions offer.

This part III includes:

- **UN GHG protocol for Companies** as a background for visualizing the full value chain of a company and its products.
- **Materiality Assessment Tool** to prioritize initiatives based on ESG/SDG impact and on business risk and potentials.
- a **tool for energy optimization and transition to Renewable Energy in scope 1+2**. This tool is a detailed process description on how companies can manage their energy optimization and installation of renewable energy, including some experience from major energy optimization projects. Here is also included a list of parameters that should be consider when building sustainable.
- a **method to transform to circular business models preparing for the Circular Economy (scope 3)**. The method is a well-tested method on how to drive the change in the full value chain, development of a new circular business model and development and redesign of products for circular material loops.
- **A framework for a Sustainability Roadmap** to manage the targets, actions, resources, and policies. This Roadmap comply with the legislative EU requirements (ESRS) and helps companies manage and communicate the strategic changes needed.

© The Author(s), under exclusive license to Springer Nature Switzerland AG 2024
G. Haar, *The Great Transition to a Green and Circular Economy*,
https://doi.org/10.1007/978-3-031-49658-5_13

The UN developed a GHG protocol dividing the climate impact of companies into three scopes, which has become the standard that companies uses and the link to the reporting of impacts according to EU legislation stated in the Corporate Sustainability Reporting Directive (CSRD). The protocol is based on the company's full value chain and is very useful in the company work on climate impact and GHG footprint, and it generally provides a good overview of all the company's impacts in the full value chain, not only climate impact.

The three scopes understanding has become the general basis when assessing all footprints of a company and understanding the origins of the impact. In principle, scope 1 is easiest, then scope 2 is a little harder, and typically, scope 3 is the most difficult. But scope 3 is often also the one leaving the largest footprints (Fig. 13.1).

The protocol has its own webpage: www.ghgprotocol.org (UN, WRI & WBCSD, 2015). This website provides lots of good material, background, descriptions, online training, and tools for corporations and cities. The material also describes reporting on GHG for larger corporations. The material is very extensive, and some companies simply do not have the resources to examine all these materials. Now, companies should rather direct the efforts towards the EU legislation on reporting (CSRD), the EU taxonomy and the reporting standards stated in the European Sustainability Reporting Standard (ESRS) that will fall in place as of 2024, with detailed descriptions on what to report on and how to report described by the organization EFRAG (EFRAG.org).

13 Introduction to Part III

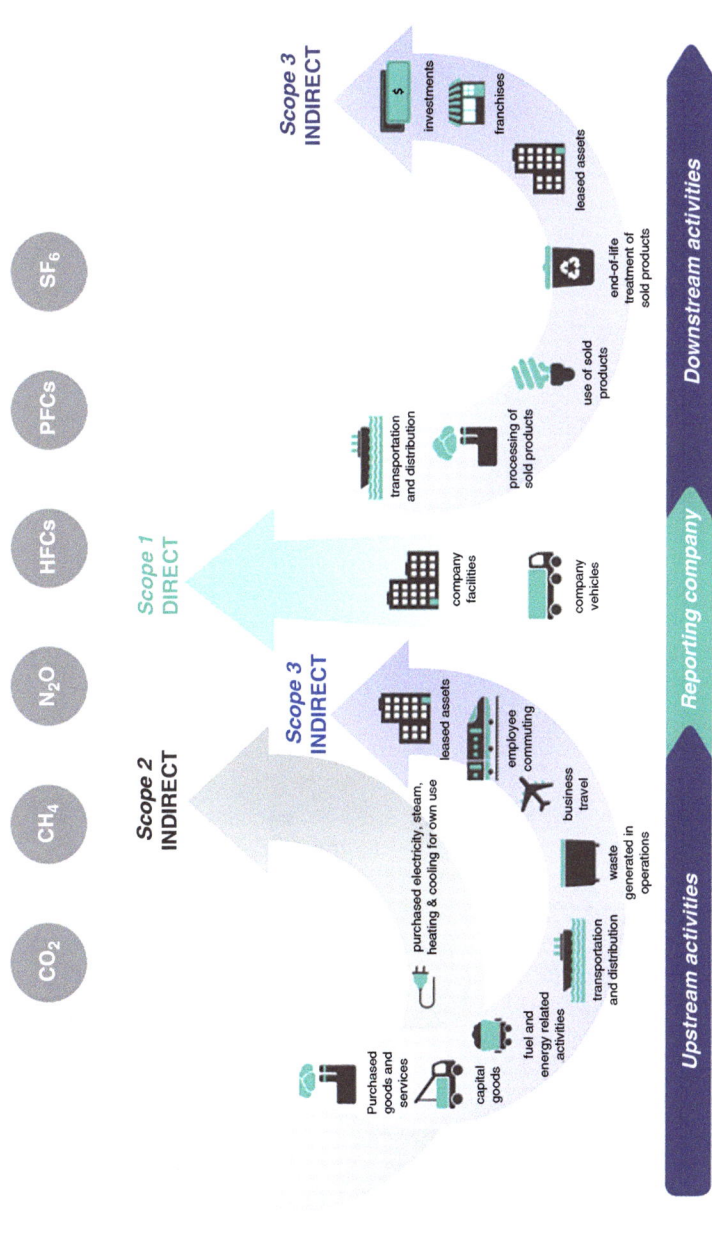

Fig. 13.1 UN Climate Protocol for companies

> **The 3 scopes are:**
> *Scope 1* covers the impact from own buildings, vehicles, and other installations owned, leased, and managed by the company. It also includes the impact from transport between own factories or sites, even if they are carried out by an external logistics partner. *Scope 1* is impacted by the choice of installations, buildings, vehicles, etc. This means that the obvious changes are made by energy optimization and implementation of Renewable Energy. Here the company choses the type of energy supply: oil-fired boilers, solar panels, or heat pumps, and whether to invest in or lease electric or fueled vehicles. So, **scope 1** covers all the consuming activities, buildings, and installations that the company can optimize within their own physical premises and on their own locations.
> *Scope 2* cover the power supply to the company, as heat, gas, diesel, and electricity purchased. It also covers potential or existing RE installations. **Scope 1 and 2** are very much tied together.
> *Scope 3* typically requires the most effort and is where the largest impacts are when assessing the full value chain. It covers the products, services, and materials upstream and downstream. Minimizing **scope 3** impacts is done by sustainable procurement, development of new sustainable products, and the development of new circular business models with low or no impact in the full value chain. It covers the carbon footprint from extraction of virgin resources through vendor and all the way to costumer and their disposal of goods. It also covers external transport of goods to and from the company, that are not done by own vehicles and transportation. The GHG-protocol web also includes a scope 3 evaluator tool, online.

Most companies start with scope 1 and 2 because here they hold the direct control of the impact and can more easily reduce greenhouse gas emissions. Energy optimization is often profitable and manageable for a start, and a good set off for the later work in scope 3. The first step for companies is to set targets in scope 1+2, and get their own "house" in order, because it makes sense and is what is expected by customers and other stakeholders.

The basis for conducting energy optimization is to define the baseline, also needed for the later follow-up on energy cost savings and the company GHG reductions. The company should include their reporting on climate impact and on climate actions in their monthly management reports in the same manner as the financial reporting. In this way, progress is ensured, as well as full focus from management to succeed with the targets and actions set. Detailed procedures and descriptions of how companies perform energy optimization in scope 1 and 2 later in Chap. 13.

> The CFO or the facility manager must be the project owner of the work on climate impact. The CFO must report on ESG targets and their financial impacts in the monthly management reports.

Value Chain Understanding to Work in Scope 3

Understanding the value chain provides a good overview and is important as the first step in the sustainability work. The company should start out by drawing the company's full value chain in a detailed and thorough way. This may be the most important step to understand the impacts fully and must include a description back to where spare parts and raw materials come from and their journey all the way through the value chain and through the company processes to where they end up with consumers and at disposal in the end. Only then is it possible to create an overview of the environmental and social impacts in all three scopes.

For climate impact work it is also possible from this to make a rough estimation of GHG-emissions. The new reporting regulations also take basis in value chain descriptions and the ESRS-standards will provide standardized value chains for various industries. It is important to ensure that the standardized value chains correspond to the actual value chain of the company, only then the reporting will become correct, and a good basis is created to set the right targets and actions to drive change.

Understanding and visualizing the full value chain of the products and services is also beneficial in conducting the risk assessment not only on sustainability but also in the assessment of future access to critical materials and securing supply from long unstable value chains. The world is changing and the fight on materials is very present, and companies are already experiencing large problems in securing production and supply. This will only increase in the future together with changing market conditions towards a Green and Circular Economy the full understanding of the value chain is crucial. The companies that experienced delivery and customer challenges during the corona times, will know that it has become even more relevant to carry out an overall risk assessment in the larger perspective of company existence.

Sustainability work is activity-based work, as we know it from activity-based-costing (ABC). Every activity in a company triggers a sustainable impact—small or large—positive or negative. Therefore, the best way to map and organize company sustainability work is to base it on company activities in the concept of value chain thinking. Then select which activities need to be changed to minimize footprints. Many companies struggle to figure out how and where to start. Here the value chain visualization and description of critical activities upstream and downstream is a good start, and a clearly visualized format will be useful later in the communication and education of the Sustainable Roadmap of the company. Earlier, this book provides inspiration on value chains from the fiber industry (paper and cardboard), textile industry, and construction industry.

Reference

UN. (2015). *WRI & WBCSD*. Retrieved from Green House Gas Protocol: https://ghgprotocol.org

Chapter 14
Materiality Assessment

The Materiality Assessment (MA) is a good tool for prioritizing company efforts on sustainability to give an overview of the impacts in the full value chain and what the efforts needed to mitigate these. The MA is the entry point for complying with the EU-regulation on reporting (CSRD/ESRS) and should be used actively as a management tool to understand and prioritize the changes that companies are to make on the impacts they cause. The Materiality Assessment (MA) must be prepared from a description of the full value chain and the impacts here, divided into scope 1, 2 and 3. Especially the efforts in scope 3 may be challenging to identify and prioritize.

Many different versions of Materiality Assessments are available online, also based on the UN climate protocol and the three scopes. Global Reporting Initiative (GRI) has developed a Materiality Assessment tool that maps impact (environmental, social, or economic) against stakeholder awareness. See: https://www.globalreporting.org. The MA from GRI was prepared as a mapping tool mainly for communication purposes and stakeholder management.

Sustainability, climate change and ESG is becoming strategic and businesses rather than communication and compliance only. Most companies today need a much higher prioritized and strategic starting point other than shaping their communication to meet the new market conditions and to comply with the extensive EU legislation. This means proactively and intensive work with environment, social and governance impacts in the full value chain and contributing to regenerating ecosystems by developing regenerative business models.

The Materiality Assessment is very much about getting a grip on orders of magnitude and should be prepared before making accurate impacts measures, thus MA is part of the preliminary work to get ready to develop goals, targets policies and actions. This is also the basis for the new ESRS-reporting directive that will be implemented from 2024 to 2026 for companies, stating: first value chain description, then materiality assessment, before measuring and reporting. The competencies needed to carry out the Materiality Assessment are often already in the company

© The Author(s), under exclusive license to Springer Nature
Switzerland AG 2024
G. Haar, *The Great Transition to a Green and Circular Economy*,
https://doi.org/10.1007/978-3-031-49658-5_14

and should be conducted by senior employees and management from production, procurement, finance (controllers), and Research & Development (depending on the size of the company), together with a sustainability expert. This work requires attention from employees having the overview of company processes end-to-end, and who have access to overall data on material purchases, sales statistics, consumption, and the like. If this work is anchored close to management and carried out at a management workshop with the assistance from a sustainability expert, it is an affordable task to prepare a good overall Materiality Assessment (MA).

According to EU reporting legislation the MA must be a double-materiality-assessment meaning that it must assess both the ESG impacts of the company and the financial impacts of the company within the ESG parameters. This is a full business assessment as ESG impact come from all business activities. Therefore, it is a good idea to prioritize which of the company's products and input materials are significant in a business perspective. Using the well-known 80/20 rule is also a good approach here, as 20% of the product typically covers 80% of revenue and 20% of input by number typically covers 80% of the costs. Thus, it is possible to get the most effect towards a sustainable business with the least possible effort. Start with the actions that creates the significant results rapid to ensure some good wins, and it's fun to see results quickly.

The MA must be done on all types of impacts and not only on climate impacts. All other sustainability (ESG) topics must be included according to the EU taxonomy. Companies impact the society and the planet in many ways, such as chemical impacts, consumption of critical raw materials and their impacts, water consumption, as well as the other environmental or social impacts stated in the taxonomy throughout the value chain. Sometimes several runs on the Materiality Assessments are needed on the various impact parameters to get a full picture of the impacts. When working with the MA, it becomes clear that there are often interconnections between different impact parameters. The company may achieve several benefits from one action if approached in a holistic way, e.g., there is a strong interaction between Circular Economy, and climate impact and again with chemical impacts.

Don't break your neck collecting a lot of detailed data before performing the first version of MA. An overall estimate is often enough and will give an excellent picture of which activities and actions can drive the changes and create the green transition of the company. A detailed mapping of the impacts in scope 3 requires professional insight. Sometimes these competences are in the company, sometimes it requires the involvement of external experts, and that effort should be saved to support the work when the actions have been prioritized. Often it will be necessary to update the value chain and revisit the impacts as new knowledge appear on for instance chemical or biodiversity impacts.

Another option is to have a life cycle analyses (LCA) prepared of the products in their full value chain on the relevant impacts, and not just GHG. It is a much more comprehensive exercise, and this effort should also be saved until the company needs to document the sustainability of new products for updated marketing and customer interactions.

14 Materiality Assessment

It is useful also to include a description and assessment of the human resources needed to implement to actions in the Materiality Assessment. Meaning what it requires in hours, costs, and investments to reach the targets set. This should be held against the gains expected. In this way, it becomes clear how to prioritize all the potential targets and actions. The Materiality Assessment is a strategic tool for prioritizing and may look like illustrated in Fig. 14.1, where both ESG impact, business impact and resources needed is included.

The MA illustrated in Fig. 14.1 is an adapted version of the Materiality Assessment and is based on the double materiality assessment added some element to become a useful business tool. This has proven very suitable as a working tool to drive change together with company management, and which is more accessible than the large material provided by GRI, especially to medium-sized and smaller enterprises. The Value Chain descriptions and the Materiality Assessment that will be provided in the Sustainability standards (EFRAG.org) are not available at this hour but are set to be published during 2023. The ESRS standard prescribes a double materiality assessment covering ESG impact versus financial impacts (risks) of the company. The tool illustrated here includes more parameters that may be useful, but the two dimensions that ESRS prescribes are sufficiently for meeting legislative requirements. It may be beneficial to include the level of effort and investments required to change the mapped activities and impact issues. A modified MA with four dimensions has proven a good tool:

Impact on environment and social impact correlate with future business potentials and risks in the sense that something that has a large environmental impact will also embed a large business risk and business potential by changing.

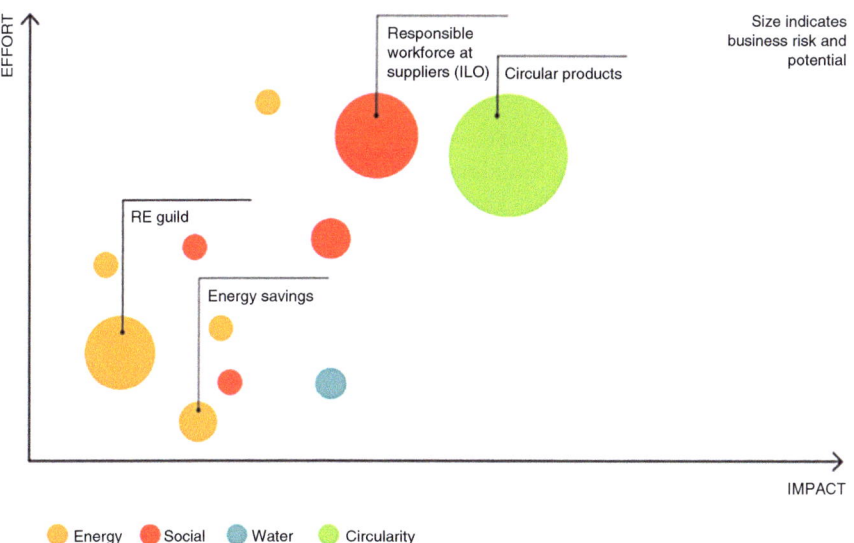

Fig. 14.1 Materiality Assessment

> **The Materiality Assessment Tool is four-dimensional and gives an overview of:**
> - The Effort necessary to drive change and reaching targets on the specific ESG parameters (calculated in time, costs and estimated invetments)
> - The materiality of the ESG element
> - The materiality of business risk and potential-indicated by the size of the circles.
> - Type of ESG parameter (or SDG) indicated by colors or other annotations, as SDG#.

The impact parameters illustrated by colors can also be linked to the relevant Sustainable Development Goals (SDG) that then becomes the strategic goals of the company. Some companies have or want to link their impacts and actions to the SDGs, as in an SDG Roadmap. An SDG map containing all or many of the SDGs will rarely create the same overview as the MA, because 17 topics is a lot to relate to. On top of that, not all SDGs are relevant to companies. Therefore, the company should first prioritize the relevant SDGs before starting the detailed mapping work. Proven tools for prioritizing the SDGs in an SDG-Strategy-House and building a Sustainable Roadmap (SDG Roadmap) are available later in this book.

This MA exercise is a management exercise, and it is very important that management drives the decision making here, often also including the non-executive board. This is the only way that sustainability will become a business driver and a genuine preparation for the new green economy. When decisions are made, the selected activities should be pursued in a traditional project management framework. This is described in detail later in this part III introducing the Sustainability Roadmap.

> **Summary of Company ESG workflow as required in European Sustainability Reporting Standard (ESRS)**
> 1. Understand the UN GHG Protocol and the three scopes.
> 2. Develop an illustration of the company's full value chain and evaluate towards the standard value chain per industry code (NACE).
> 3. Prepare a Materiality Assessment with overall estimates of the company's impacts in both scope 1+2 and scope 3—see later this book.
> 4. Build the ESG report, starting with a baseline for scope 1 and 2 and initiate actions here, as energy optimization, etc.—see later this book. Start early enough to have internal versions.
> 5. Management's prioritization of the company's targets, actions based on the MA.
> 6. Allocate resources and involve experts as early in the process as needed.

7. Appoint responsible project owners in management, that can appoint project leaders. Remember CPO and CFO are the most important project owners.
8. Integrate into the business and prepare a Sustainable (ESG/SDG) Roadmap as an implementation plan.
9. Integrate into management report and follow up every month, and at executive board meetings and board meetings.
10. Develop education and communication plans for internal and external use.
11. Build the mitigation tool for handling ESG situation addressed by employees and management.

Chapter 15
Changing Company Climate Impacts in Scope 1+2

Once the company has prioritized its ESG targets and actions it must start the concrete work. Most companies start with climate impact reductions in scope 1 and 2, meaning energy optimization and transition to renewable energy supply on their own premises. This chapter provides methods and recommendations for companies in their work on minimizing climate impact in scope 1 and 2. Methods on scope 3 are provided in a later chapter.

Energy optimization is often a profitable investment in reducing company costs. Often it is more about common sense and normal cost awareness than climate action. A management that has NOT done the business case on energy savings and potential investments in RE deserves to be called cost-irresponsible rather than not yet focused on sustainability and climate change.

The European energy-saving scheme to meet the CO_2-quota of the Kyoto agreement was introduced in the EU in 2005. It has promoted energy savings in many countries even though it seems as if there still is a huge potential to be harvested. In 2019 the EU agreed on a target of 32.5% savings in 2030. This scheme ended in 2020, and new programs are introduced to drive energy optimization and electrification of the industry away from fossil fuels. The old scheme resulted in some major projects around Europe as well here in the Danish municipalities and companies and showed that up to 40% of the energy consumption could be removed based on a profitable business case. Below are some cases and experiences on an overall level.

Since then, the war in Ukraine and EU dependency of Russian gas became clear with increasing energy prices and fear of lack of supply. It is now obvious that energy is geopolitics and EU has set the goal to be independent of imported gas from Russia before 2030. This has changed the energy situation in EU and political power has been put behind the transition towards independence of imported energy rather than of fossil energy. We see now that extraction of local gas is prioritized higher than climate impact, and nuclear power and biogas are defined as *"sustainable"* in the EU taxonomy. This seems a little unscientific and may be changed when the energy situation in EU is more stable. The increasing prices of energy has made the business cases of energy optimization attractive. At the same time, the price of renewable energy has also decreased, and are now compatible to all fossil fuels, also coal. It is worth for companies to look at investments in RE.

The 11-step process is based on learnings from large energy optimization projects in Denmark that along with some takeaways made these projects successful. Both in the involvement of top management and in reaching ambitious reduction targets (Haar 2021) (Fig. 15.1).

1. Organize this as a *cost-cutting project*, not a sustainability project. The project owner is the CFO, and technical and financial competencies from the company should be involved, to drive the change needed and to maintain the reductions in the future.

2. The best *projects include all relevant technologies in one scenario* and has a *holistic approach*. It should include insulation, ventilation, exchange of light sources to LED, and energy-producing machines that need cooling. The best financial results, climate protection and indoor climate will come from monitoring and evaluating interaction from alle technologies and all energy consuming installations and effects hereof.

Fig. 15.1 Process for minimizing GHG emissions in scope 1 and 2

3. *Build a baseline* of a full year's consumption end emissions before starting to optimize. The baseline should include both the energy consumption data, CO_2 emissions and energy costs. Typically, these data are kept in different systems and do not reconcile. This makes project management and follow-up on cost-savings difficult afterwards. If data are not reconciled before start of the project. Companies risk not to harvest the estimated cost reduction e.g., if the project is based purely on climate data.

4. Use *external, independent advisors/specialists* to prepare a report to uncover the potential savings, then challenge them to reach an extra 10%-points. Make them challenge the solutions and the vendors. Be careful with advisors that have other incentives than your reductions like ESCo, advisors from electricity providers, or on public schemes paid by funds. They may not go all the way for you. Make sure there is no kickback involved to providers or vendors—or at least that the pricing here is transparent.

5. Make sure that the specialists also include *Good Indoor Climate* as a goal—maybe even involve an architect. You do not want unhappy employees or bad light – that is much more expensive than the energy you save. Often you can get energy savings and good indoor climate at the same time if you use good advisors with a holistic approach and the right experiences.

6. *Include electrification of production* as this itself causes energy savings, since the combustion engines are much less energy efficient than the electrical driven machines. It is often large investments also causing retrofitting of the plant site and involves experts and vendors specialized for this, but it is a preparation for the future, and it will make the installment of RE even more profitable as they are typically electrical power producing.

7. *Build a real business case* compiling all technologies and initiatives and do not use simple payback time. Use ROI and include the lifetime of the installations, RE, and changes suggested. Then the low hanging fruits will pay for your long-term investments as RE. Ask for more than one optimization scenario—e.g., 25%, 30 and 40% reduction. Remember that all initiatives with a ROI above the interest rate on a potential loan is a win for the company.

8. *Celebrate the successes* and the changes achieved for the climate and for the employees—it makes people proud to save money and avoid climate change at the same time.

9. *Measure* before you start, *measure* all the way through and continue to *measure* afterwards and maybe even compete internally or benchmark externally—then you will also harvest and maintain the savings from behavioral changes not only from the hardware installations.

Remember that changing installations, windows and so forth also costs GHG—*think circular* when you replace stuff. Calculate the carbon footprint of the construction material you buy and use LCA including scope 1 + 2 +3 as the basis for your GHG emissions.

 *Do energy saving before you install RE*—the cheapest and cleanest energy is the one not used.

Experiences and Best Practice

Here experiences for the actual execution of energy optimization and installation of renewable energy are shared.

Baseline

The calculation of the impact from scope 1 and 2 is simple. Data is already in the company's financial records from the bills of consumption by the utility provider. In many companies, often only the energy costs are recorded, but behind this are data on energy consumption, fuel and other things in appendices and statements from the utility provider (electricity, district heating, gas, oil, diesel, petrol, mileage, etc.). The finance department should be able to provide the data for preparing the baseline on the historical consumption, which will be the basis for measuring the future gains. There are lots of templates available online to assist in calculated the yearly GHG emissions in scope 1 + 2. The simple ones are even free of charge, but the more complex version covering larger corporations and other types of ESG data are not. Examples of ESG dashboard are available providing tools for driving change including the Materiality Assessment and Sustainability Roadmap. The most recent full year is typically chosen as the baseline year and should be the year before any actions are taken. The baseline should cover a full year to offset seasonality and fluctuations in the weather (degree days).

Conversion from consumption unit (kWh, liter, m^3, etc.) to CO_{2e} is done with a conversion factor provided by the utility company. For the climate impact from electricity consumption, there is often a national report on the share of renewable energy in the power supply on website of electricity provider. For passenger and goods transport in own vessels and between own sites, there are published conversion factors from various sources as; for example, EU8, UN, or other official sources. Car brands and logistics providers also report emissions per kilometer/mile, or the total consumption in CO_{2e} over a distance. It's a good idea to use official third-party assessed conversion factors, as both car producers and logistics providers tend to be very optimistic and underestimate the GHG.

Technical Learnings

The business case is typically better for liberal professions, offices, private homes, and public operating buildings than for industry, as industry often is refunded a larger share of the energy tax. On the other hand, investments in renewable energy technologies are now so profitable that they are worth a serious consideration. Renewable energy plants hold a lifespan of 30-40 years and can supply electricity for many years, so it is a bad idea to use the traditional parameters such as simple repayment time. Renewable Energy (RE) is the future and are more economically sustainable than manufacturing technologies. RE are not a production plant, it is a power supply plant, and can be traded separately.

> **Energy optimization in buildings and manufacturing plants fall into four categories:**
> A. Low hanging fruits with a very attractive business case, like balancing of ventilation, installation of LED, sunscreens on windows, etc.
> B. Retrofitting with longer lifespan and payback time, like insulation, energy glass in windows or removable screens for insulation, etc.
> C. Change of energy source, electrification, and installation of Renewable Energy
> D. Behavioral changes

The first three categories are retrofitting of installations and building envelope, whereas the last one comes from behavioral changes, difficult to achieve and even more difficult to maintain. More words and descriptions are put to these four categories here.

A) *The low hanging fruits* are typically removal of the passive consumption (stand by) of energy when buildings or machines are not active. That goes for management and maintenance of ventilation systems, automatic light control, automatic control of standby positions of machines, etc. It is a good idea to consider an energy management system to monitor energy consumption and map it to the actual use of buildings and plants.

   The passive use of energy (stand by consumption) is often much higher than expected and very easy to eliminate. Next is the maintenance and management of the ventilation system, A/C systems, and heat pumps. These technologies are very energy consuming and are also a very easy way to save energy—make use of experts and remember the focus on indoor climate—a badly managed ventilation and A/C system often means bad indoor climate and a potential for energy savings.

   The exchange of light sources to LED is also a low hanging fruit because the BC is so attractive. Be sure to require the needed light intensity and not to waste our good, old fittings replacing them with cheap single-use LED fittings that need full replacement after a few years. There is a lot of cheap stuff on the market and some of this is not even safety approved. Your old light fittings can be retrofitted and if a replacement is necessary, make to acquire fittings with a separate LED source that is replaceable.

Shielding windows from strong sunlight with either awnings or film directly on the glass is also a low hanging fruit that will improve the indoor climate.

B) *The retrofitting* that requires larger investments is the insulation of envelope, windows, and doors. Replacement of windows and doors is often not profitable unless very cheap materials are chosen and then the lifespan is hazarded, and the climate emissions from such a linear product will more than equalize that of the energy savings. Look at potentials for refurbishing and insulate windows, doors, and roofs. Be very careful when insulating walls because that might cause mold if not done correctly, making the building useless. Remember that warm air moves upwards to the roof, and the insulation of roof is much more efficient and typically also easier that insulation of walls. Refurbishment and insulation of existing buildings is often the best climate solution even if you do not save as much energy from consumption, as when buildings are new. Looking at the full value chain (lifecycle) of buildings it will be the most climate friendly solutions to renovate and retrofit.

Especially if the building is made from bricks and of older age, make the best of the good old materials and do reasonable insulation instead of replacing many good quality materials with minimal benefits. Here it is important that you use an independent good advisor not to risk replacements for inadequate reasons. The new building regulation must now take the full lifecycle and lifespan into account when climate impact and energy consumption is calculated.

C) *Replacement of energy source* from fossil based to Renewable Energy becomes very financially attractive due to the decreasing prices of RE and the increasing prices of gas and coal. Especially if it is done on the right scale and balanced with energy consumption. Different approaches apply for companies:

- *RE-Guild* providing energy directly or indirectly to companies in a specific area or within a group of companies in an industrial urban area.
- Directly requires a separate grid, storage, and balancing of energy according to consumption.
- Indirectly the RE installations are connected to the national//regional grid and are managed and balanced here. Shareholders of the guild sell the power to the grid owner or the state. Some countries will offer benefits or periods of fixed prices for electricity to promote investments in RE. Profit from the RE-Guild is then paid back to shareholders or invested in additional installments.
- *Company held RE installations* balanced to provide the exact consumption of a specific company. If the installations are off grid, it will require storage and balancing systems. In some countries, like Denmark, this type is not allowed because a common control of the SmartGrid system will provide a much more efficient energy and secure supply. In other regions with poor grid infrastructure and low security of supply, this is very recommendable if the plant facilities are of a certain size and the energy consumption is significant.

We see more and more production sites being provided privately and locally with RE in rural areas—giving a more stable supply of green energy at attractive prices. It, of course, requires competences to maintain and operate these installations, but that is manageable.

- *Electrification* of manufacturing plants and equipment is an important part of minimizing GHG emissions. It is also energy efficient because the electric engines are more efficient that the combustion engine. As renewable energy often comes as electricity it makes very much sense to shift to electrified production. It is not an easy task and requires not only a lot of technical competence, but also investment in technology. Some manufacturing processes are not easily transformed to electric input, but the potentials are still enormous because it both provides energy savings and cheaper input as electricity produced from RE will become the normal where prices are controllable on a regional level in the future. It will almost always require involvement of technical experts from outside the company and extensive planning of the transformation with product line stop and replacement.
- *Compensation through credits schemes* where companies buy certificates of green energy installed elsewhere (Kyoto quota system). This may be a solution for smaller companies, or companies that do not have space for RE installations themselves. Make sure that *additionality* of RE is guaranteed. Some providers issue certificates based on already state or EU financed RE plants and that is cheating a little because they would provide green energy to the grids anyway. Auditors even approve this kind of certificates. Remember that an electron is an electron, and nobody can see which source it comes from when it comes out of the plug. *Additionality* is ensured when power providers spend the extra fee from certificates to additional, direct investments in RE that would not have been invested in without this type of certificates. The latter are typically a little more expensive than those from state-owned plants are. Make sure that the certificates are genuine, audited, and traceable to actual installations with access to financial statements and accounts on power produced. Fraud is out there.

    A new type of credit schemes is coming for compensation through investment in forests—existing forest or new planted forest. Compensation across the Climate Nexus is challenging, and it can be argued that this is not the way we create climate action. It is a good idea with a lot of forest and EU has a target of three billion new trees in 2030, but this does not compensate for emissions from fossil fuels. Minimizing and optimizing energy consumption and investing in Renewable Energy is the way to climate actions from companies emitting GHG from fossil fuels. THEN it is a good idea to invest in forest and wild nature.

In the future we will see much more RE installations in different constellations as also shown by the fantastic story of the America Military. They have been ahead of their transition program, ever since they found out that RE was much safer for the soldiers since they could hear the enemy sneaking in on them; and the energy source cannot be cut off as it could when it came in trucks with diesel. The world would have looked a lot different if we had chosen the electric engine and RE 100 years ago, as Nikola Tesla suggested. Below are included a few successful cases for inspiration that were built on the above method and recommendations.

D) *Consumers and employees' behavioral changes* in the use of energy is an important part of creating and maintaining energy savings: Although this is often the hardest part to achieve. There is great potential in reducing power from standby modes for machines and the like. Old habits and chords are in the way of focusing an energy savings from behavior, as turning off the lights in rooms that are not used, is still a challenge. This can now be done by sensor technologies and other monitoring technologies, which make it possible to get rid of unnecessary power consumption in easy ways. Energy management systems also help to monitor the company's use of the power-consuming devices, and some systems offer monitoring down to individual groups or areas of production. Competitions and benchmarking are valuable tools in changing employee behavior systematically. Many energy advisors often do not include the behavioral changes because it is difficult to measure, maintain, and follow up. However, the potential here should not be underestimated, and examples where large savings can be harvested are typically made by involving employees, as: server rooms, older machine lines, commercial kitchens and warehouse and production lighting.

The work on optimizing energy consumption and switching to renewable sources is a continuous work and monitoring energy consumption on a detailed level is important to maintain and identify new potentials all the time. Some cases for inspiration are shared here. When starting out the energy optimization it is a good idea to perform a feasibility study based on pilot installations or pilot buildings if the company is a large asset owner.

Short Cases for Inspiration

A couple of cases are included here, and, in both cases, significant savings were estimated.

- A large HQ office building in Copenhagen energy optimization at an investment of € 374,000 resulted in a yearly power saving of € 413,000 accounting for 36% of the total power consumption and a better indoor climate that was notable by employees.
- A large municipality in Denmark raised the GHG reduction target from 25% to 40%, solely because of the financial potential and the uncovered technical possibilities in several pilot buildings. The municipality could obtain loans for energy optimization at an interest rate of 4% per annum and thus everything with an IRI above 4% over a period 10 years (technical lifespan) was profitable and an income for the municipal increasing welfare.

In both cases, significant savings were identified in the technical feasibility studies. In the office building, the proposed changes were implemented resulting in a better indoor climate for employees. In the municipality, the original energy optimization project has expanded to a larger climate initiative: GoGreenwithAarhus.dk, which shows an example of a climate-friendly municipality. Green investments of DKK 450 million in energy savings in the municipality's own buildings (project Aa+) until 2019 and the investment is expected to be paid back to the municipal over the next 15 years.

Typically, industry plant owners will be able to cut between 10-30% off their energy costs and GHG emissions. Again, it is poor management not to reach for the energy savings—it is not just lack of sustainability and climate action.

Sustainable Buildings in Scope 1 + 2 + 3

Sustainability certifications of buildings (the entire construction) are spreading rapidly, and many of the large asset owners and property investors only invest in certified buildings.

The typical sustainability certifications, all of which require third-party verification, are:

- LEED (USA), Leadership in Energy and Environmental Design, which focuses on sustainable environment, efficient use of water, energy and atmosphere, materials, and resources, and on indoor climate and environment.
- BREEAM, the UK Building Research Establishment's Environmental Assessment Method, which is an LCA-based method with the primary focus on health and environment-related aspects, such as emissions from construction products, waste management, resource efficiency, ecology, biodiversity, climate adaptation and partly on social aspects (e.g., durability of design, possibility of transport).
- DGNB, is a German ecolabel and is becoming more and more widespread in the Northern part of Europe including Denmark and focuses on sociocultural and functional quality, environmental quality and economic quality and is a benchmarking tool. DGNB is also starting to integrate LCA as a method.

The certification of a building under one of the schemes will require that the building materials carry an ecolabel, such as the EPD, Cradle2Cradle, or others, for documentation and third-party verification. Some in the industry are already preparing for the future EU-PEF and DPP.

The elements that should be included in a sustainable construction, regardless of certification or not, are the five elements in Fig. 15.2 reaching towards an Impact Positive Building.

The climate impact of the building based on the principles of LCAs has two legs: (a) the construction of building (embedded CO_2); and (b) the energy consumption in the use phase. For many years the focus has been only on the energy consumption of the operating building. But it becomes clear that more than half of the climate impact in the lifecycle of a building comes from the building materials, their manufacture and disposal. A recent Danish study

Fig. 15.2 Sustainable Buildings-creating positive impacts

showed that 70% of a building's climate impact occurs before the building is operating. Legislation is switching towards the principles of LCA and therefore, both measures and impacts are to be considered:

- a circular building that reduces the impact from building materials and
- an energy-efficient building with low energy consumption, possibly with renewable energy solutions integrated into the building that will become energy positive in the future.

It is also important to set requirements for the indoor climate in the buildings and include it in the assessment of the building. The building must be assessed on energy performance and indoor climate in operation and the pricing and guarantee of the building must be dependent on this. See more on *Commissioning of buildings*, later.

Circular construction describes how the building itself is constructed, of which materials, and the disposal of the materials at end of use. A circular building contains of materials that are both recycled and recyclable in the future. Some building materials have a very large climate footprint, despite having a circular potential, as for example concrete, bricks, and tiles. Bricks and tiles have a large climate footprint when manufactured, but these materials have a good potential for being recycled. Therefore, these materials must only be used when they are recycled, and in ways that they are recyclable again when the building is to be renovated or demolished in many years.

Wooden structures have a significantly less climate footprint than virtually all other materials, and today there is plenty of science that proves that wooden buildings are just as fireproof and safe as other buildings. Often, wooden buildings will have a better indoor climate than most other buildings. A circular building must hold a potential for long durability and maintenance. Therefore, it is important to set requirements for the building's ability to maintain and repair.

Today, many building materials on the market are sold as maintenance-free, but they are often linear products that will end up as waste after a short lifespan because they do not hold the potential for repair, reuse, or recycling. Coloring or coating roofs white should also be considered, as it has a direct positive effect on climate change.

C *Energy-efficient construction with renewable energy.* Energy efficiency is becoming a prerequisite in building regulations all over Europe, and it is obvious to integrate renewable energy into building construction, as solar panels, wind turbines, heat pumps, ventilation, recycling, and solar collectors (hot water). Most important is to take a holistic design view of the building, its consumption, and energy supply. It is always cheaper to integrate renewable energy from the design phase than to add it on later, and often also prettier. Today, there are many interesting integrable renewable energy solutions which also have beautiful architectural value. In addition, rainwater resources should be utilized in and around the building for toilet and washing (not shower), which today are found to be cost-effective.

D *Wild nature and regenerative ecosystems.* Decline in biodiversity is one of the major challenges, and building owners must consider the potential of creating biodiversity locally. It can be anything from a more passive wild flora in the grasslands, which requires minimal care, to active biodiversity, which also has recreational value for nature and employees, such as beehives, utility gardens, insect hotels, etc. This typically entails a rather limited expense, especially if it is included from the start, but it has a huge value for employees and for the signal that a domicile sends to the outside world.

More and more municipalities are demanding biodiversity and green roofs in new buildings. The new EU-taxonomy in the Sustainable Investment Directive (SID) includes requirements on regenerative nature and biodiversity and buildings are covered here. The new EU-biodiversity directives and global commitments on biodiversity also effect the building and construction industry, as the impact on nature from mining for the industry is large. The legislation requires full responsibility of the full value chain and the impacts from building materials, mining, etc. on nature and biodiversity must be accounted for when claiming sustainable construction.

Wood as a construction material is achieving special attention since it embeds carbon and meets the climate targets. It is much needed to innovate when designing buildings, but it is also important to reuse and recycle the materials already available. Harvesting virgin wood for construction may embed carbon, but the industrialized forest industry causes other problems as lack of biodiversity and wild nature. A commercial forest is almost a monoculture as known from agriculture and a shift to wooden constructions will increase demand for virgin logging. If the wood can come from recycled wood and if material loops for reuse and recycling of wooden products and materials are created, the shift towards more wooden constructions is reasonable. Today recycling of wood for construction is poor and most recycling of wood is downcycling, and not circular. Material loops for other construction materials made from metals and glass are more developed and the environmental impacts from these are typically lower than constructing from new logged wood.

Climate adaptation has become necessary since climate change is a reality. All around the world we experience torrential rains, flooding, burning of woods and grassland, and rising water levels, also groundwater levels. Companies need to protect their buildings and their assets from these changing weather conditions and especially floodings. This is done by draining the water, collecting rainwater that can be used actively in the building, aboveground drains, fascines, and water collection that can be used in watering the new biozones to increase biodiversity. In addition, important installations and electrical panels should be secured by placing them high enough to not be affected by floodings.

Sustainable and certified buildings have a proven higher sales price, higher rental price, cheaper insurance and better conditions on loan and mortgage. It is important that the business case for the building is set up for the entire lifetime of the building, i.e., the Total Costs of Ownership. The business case must also include non-financial ESG data such as climate impact, climate adaptation, circularity, biodiversity etc. Only then the calculations on future financial returns are valid.

Commissioning of Buildings

Requirements on commissioning (continuous measurement) of the building's performance on all relevant parameters from the building is handed over to the owner, and continuously here after when the building is operating. Not all Building Certification Schemes require Commissioning, but it may prove to be an economic advantage to choose this. It keeps contractors and consultants up to the actual performance of the building. In some countries commissioning is part of national regulation, but in many countries documenting building performance only requires theoretical inventories on the energy efficiency of products and not the actual energy efficiency when the building is operating. Unfortunately, fraud appears and especially within the hidden installations, and this can only be revealed by continuous measurement of the building's performance. Commissioning measures should include not only energy performance but also on indoor and outdoor climate, climate adaptation, biodiversity and so forth.

Reference

Haar, G. (2021). *Climate Nexus - The climate challenges in a company perspective (danish)*. Quare.

Chapter 16
Method to Transform to a Circular Business Model (Scope 3)

This chapter provides methods and recommendations on how to approach scope 3, and a method on how to transform to a circular business model. Scope 3 impacts and GHG-emission comes from products and its full value chain and minimizing the scope 3 impacts are significantly more complex, and a much more based on the specific industry, compared to scope 1 and 2. The method on developing a circular business model here, is not only harvesting climate impact but also creating resource efficiency and regenerating biodiversity.

Scope 3 of the UN Climate Protocol covers:

- The impact of products throughout the value chain, i.e., from raw material extraction to disposal, and especially end-user consumption patterns, has a major impact on the impact of products.
- The transport of products by means of transport other than within and between own premises. This often account for a smaller proportion than many assume, but we need to transform the transport sector towards electrified and shared transport for many reasons.
- Employees' consumption from their transport, office materials and other

The largest effects achieved in scope 3 are to make the products circular and develop a circular business model that ensures infinite material loops, phasing out of virgin raw materials and waste. Recycling and recycled materials have a much lower climate impact than raw materials that are virgin extracted (2). This is why the starting point for a scope 3 analysis is to investigate the use of raw materials and/or material inputs, as well as the goods sold in quantity, quality, and price. It is typically simple layouts of data that the company already has in the purchasing department, sales department or production department that form the basis for getting an overview of scope 3.

> **The impacts in scope 3 can typically be calculated in two ways based on:**
> - Purchased materials and sold products based on the company's internal registrations and the mapping of the full value chain to trace back the full impacts of scope 3.
> - The product's life cycle analysis (LCA) based on third party analyses for the product specifically or more generic LCA estimations becoming available in the future.

Most often, the company does not hold life cycle analyses (LCA) of its products, and it is costly to have these prepared for specific products. Having a LCA developed should be saved for documenting the sustainability of future products in the lifecycle. The company should start from its own data on materials and products by compiling lists of purchased materials and goods as well as sold products for a period, for example a full year.

It is important also to include the packaging material in the scope 3 work, and for some industries packaging represents a very large share of the climate impact, while for others it is a small share. Great attention is on minimizing packaging materials and to choose sustainable packaging, why it is always an important element to consider. Packaging is a visible part of the product and creates identity, and often represents large volumes. The share of single-use packaging material has grown dramatically in recent decades. The packaging analysis must cover primary, secondary and transport packaging, and there are major financial benefits in making secondary and transport packaging reusable. Manufactures and marketeers are subject to Extended Producer Responsibility (EPR) for packing material in EU, as of 2025.

To minimize climate impact GHG-emissions per selected material type or product type throughout the value chain, and through entire material flows must be assessed. The company's own consumption data is multiplied by the conversion factors for the climate impact of the material. There are various ways to identify valid conversion factors, such as industry data from in industry associations, or a material sustainability experts can provide independent conversion factors.

Lifecycle Analysis (LCA)

In EU this will in the future be covered by the reporting requirements from ESRS that will provide recommended links to LCA databases, etc., but there is still way to go before these standards are available.

There are published life cycle analyses (LCA) from scientific studies available with useful conversion factors. They may be found online or through industry associations. It is important to evaluate the methodology and approach of the data used in these LCAs to ensure that it covers the right value chain, the right material loops, and that the parameters chosen are valid. The climate impact assessment in scope 3 should be anchored in production, procurement, finance and/or business controlling, and not with the CSR people. Then, climate and sustainability work are integrated into the daily operations from the beginning.

It requires professional assistance to have a life cycle analyses (LCA) prepared on selected products, as a second method. New LCA-consultants are on the raise, and they can perform these LCAs providing an estimate of the environmental impacts of a product. LCAs are often costly to prepare for an SME and there are yet no EU standards for the methodology or databases that provides the impacts on which the LCAs are based. This standardization comes with the new EU-PEF (Product Environment Footprint) and covers scope 3. Therefore, a good start is an approximated analyses based on own data and publicly available conversion factors. Later, the company can have specific life cycle analyses (LCA) performed as part of the documentation on new sustainable products. Some industries may be lucky being an industry where the EU has already published guidelines for the product categories of the company, and then it is easy to decide on a good solution on how to assess scope 3 (8).

Finding most non-financial data is still a detective's job, and these are data that the company may not have access to, or is not used to handle, nor formats or reporting standards has been provided yet.

It is important to included it in the Sustainability Roadmap how and when the company will uncover their impacts. A company is typically not able to calculate the entire scope 3 at once. It is an important task to set the framework for how to draw up the necessary inventories and set the necessary priorities. Thus, it is important to notice that the CSRD with the ESRS on reporting will release requirements and standards during the next years. The first reporting period for the large corporations is the Annual Report for 2024, so then there must be guidelines available, or they will need to postpone the scope 3 estimates.

Once the company has an overview and developed policies, targets, and actions, a very large part of the work in changing the impacts from scope 3 is to develop a sustainable procurement strategy that enables procurers and employees to make the right choices every time. This is also a strong communication tools toward vendors. A circular business model that ensures reuse and recycling of products and materials will become a market requirement, and a large task to implement in the company and in its future contracts.

Summary on creating data for scope 3
- Prepare lists of purchased materials and products sold for a given period—typically a year or a quarter. Arrange these lists in order of the largest quantities of materials purchased and products sold. Remember to include packaging materials on the lists.
- Form an overview of the main material flows (end-to-end) from the lists and identify the large volumes and the major impact matters. Remember the 80/20 rule and focus on materiality.

- Calculate the climate impact of the most important materials/products for CO_{2e} in the entire value chain.
- Also use the above data in the design the full value chain(s) of the company, then they become illustrative and comprehensive.
- Select the most important materials/products and prioritize the obviously low-hanging fruit as the starting point.
- Make a data collection plan for when the company will calculate the impacts from the different material flows and product streams. Or comply with the reporting requirements in EU.
- Incorporate into the Sustainability Roadmap. See later.

Developing a Circular Business Model

Many executives and business owners face the need to transform their businesses into a circular business model to ensure company competitiveness in the future. Customer demands and new legislation mean that many companies will soon face the necessity for transition.

The drivers for companies are:
- Consumer demands for more sustainable products, demands for transparency and traceability together with a rising demand for reused and recycled products. This is also embedded in the EU PEF.
- Profitable business models based on return systems and clean material loops to ensure extended lifetime of products and materials, reuse and recycling of products and materials.
- EU Green Deal—Strategy and Action Plan on Circular Economy (2020) and the following national strategies and legislation. Also demands from the North American market legislation will force companies who export overseas to meet the legislation from here.
- Increasing focus on climate change and on the fact that a very large share of GHG- emissions originates from our products, overconsumption, and disposal.
- Tenders from EU and state funds increasingly include requirements on Circular Economy.
- Virtually all soft money and public funding from the EU is dedicated to climate protection and Circular Economy (as HORIZON and the large number of SME-programs granted these years as well as EU Structural Funds).

In this chapter a process to transform into a circular business model is presented. This process is developed as a simple strategy process focusing on the full value chain of products and materials, rather than simple go-to-market strategies. It is based on years of experience in transforming company business models also resulting in the redesign of products and substitution of materials. Experiences come from years of business development, and gathered from the work with 40 companies, all of whom have designed a new, innovative (green) business model. These 40 companies participated in public programs for New, Green Business Models and Resource Efficiency financed by The Green Transition Fund (Danish Business Authority) in the years 2010 to 2015 (25). Some of the participating companies are also in the Case Collection (Haar 2024b/24).

In the work with these companies a general tendency became clear of which companies succeeded, and a certain set of criteria was in place in the companies that succeed in the transition. The most mature companies received additional support (soft money) for one year to implement the new business model and get ready for the market.

The following criteria are important for the success of transforming to Circular Economy:
- *Management holds a deep insight into Circular Economy* and understands the business potentials of a circular business model. To many it is still a new experience, that sustainability and Circular Economy is a core business discipline and not a CSR task on the side. Experience shows that if Circular Economy is anchored in communication or the CSR department, the commercial or environmental benefits are not achieved, as if managed as traditional business development and strategy.
- *Companies base their work on the competitive advantages and increasing profitability* as the reason for the transition. There must be a focus on the commercial benefits throughout the value chain and a focus on the difference in a circular value chain rather than a linear value chain.
- *Management sees this as a new business strategy* and are deeply involved in the full process. The best ambassadors for the transition are the CFO and the COO, since it is about new products, materials loop, and building a business case.
- A *traditional Business Plan* is built, including business case and implementation plan, as the driving tools—like any other strategy process.
- It is often necessary *to involve technical experts* early in the process, both in the investigation phase and in the implementing phase. The environmental impacts must be quantified in a professional manner based on scientific methods—typically Life Cycle Analysis (LCA). This is the basis for at trustworthy communication afterwards.
- Implementation in *SMEs* is often limited by human resources (hands and heads) and by funds (loan and equity). Because the transition requires investments in product development, production equipment and marketing.
- Companies *benefit from deadlines, gates, evaluation, and sparring* with external experts or a board of skillful senior executives.

The transition to Circular Economy is traditional business development in the company preparing for new market conditions. The difference is that the external circumstances and market conditions will become very different from what they have been for many years. Business models need to be fundamentally changed—both from the outside-in and from inside-out. Therefore, the green transition must live in the boardroom, and the boards must give this as much attention as Audit, Risk Management and Digitalization get today. In the future, companies will hold a sustainability committee in the non-executive board, as an audit committees and remuneration committee. The green transition will threaten the company's existence if the business models are not adapted in time.

The method in Fig. 16.1 shows the process used on transforming existing companies and is developed and tested in number of Danish companies.

Unlike many other strategy processes that are based only on customer demands, this is process it based on the products and the material loops as the starting point.

Fig. 16.1 Process for transforming to a circular business model mitigating scope 3

Developing a Circular Business Model

The ENTIRE value chain must be mapped for the products and materials. The future requires well produced and redesigned products for the Circular Economy, and companies need to know their value chains and their products far better than many do today. A company needs genuine, well-documented, circular products to market them. The days of greenwashing are over.

The process is designed for the conversion of an existing business with an existing customer base. Of course, it is important to find the unique market position and create a unique Value Proposition for the products when they are ready. The business plan must include the traditional Market Analysis, Market Positioning, Competitor-Map and Go-to- Market strategy, but first there is a need for understanding the impacts of the products. On the other hand, the green transition and the demand for circular products open the markets in new ways and provide new market positions. The steps in the model are explained here.

A. Mapping the full product linear Value Chain (AS-IS)

Companies often believe that they know the value chains of their products, but many are surprised when they start mapping the value chains with Circular Economy as a goal, to find out that they are not aware of the origin of raw materials or how the products are disposed after consumption. Companies must undertake a thorough work with the mapping of the full value chain. They need to identify if products contain critical materials, and then it is often necessary to combine several value chains both downstream and upstream.

By mapping the entire value chain, the company gets the first view on how circular loops can be established and how new take-back systems of products or materials can be established. Initially, this is a mapping of the current situation (AS-IS), and then the future circular value chain must be prepared (TO-BE), but this requires a deeper insight into the materials in a circular perspective.

The entire product value chain must be understood at the product level. A diagram is needed including potential new take-back systems or linking to existing take-back systems of products and/or materials. This will visualize which steps in the future value chain are relevant, and which are irrelevant for the company and the new business model. This process of identifying and drawing the value chain (as-is and to-be) clarifies the challenges and the future position the company in the circular value chain.

The company must keep new digital opportunities and new sales channels in mind throughout this process. It is also important to look at the changes that occur when products must live longer and are reused, and materials are to be recycled.

It is important to keep in mind that the green transition and digitalization in concert will disrupt many value chains. Part of this mapping must prepare companies to see their product value chains in new ways (Haar 2024a). As Circular Economy and EU legislation are implemented, the way materials are reused, recycled and how new return systems are established, will change. The sooner this is understood, the sooner a company can adapt. Here, it is important to involve industry specialists, and/or Circular Economy experts who have an overview of the restructuring of material loops and that can ensure that the dogmas of Circular Economy are already complied with in the mapping of the future value chain and position. Most importantly, look at the potential for recycling and extending the life of the products within the company's own discretion, so that the most valuable elements of the Circular Economy are harvested first. For inspiration, examples of old and new business models for fiber-based printed matter and plastics are shown below.

B. Understand the Value Chain and the Material Flows in a Circular Economy

Once the products existing, full value chain is mapped, an understanding of the materials streams through the value chain must be established. Companies may want to extend their stake in the value chain to ensure circularity as the responsibility of the value chain expands from scope 1+2 to scope 3. The entire material flows are from primary production or virgin mining throughout all production steps, steps of consumption, distribution, reuse, recycling or degradation of the product and the materials in a full life cycle. Industry specialists and/or Circular Economy experts have an overview of the technical potentials and challenges of the materials and can qualify the future conversion of material flows in circular loops. It is important that Circular Economy principles are incorporated at this early stage.

Materials that have a potential for reuse and recycling must be identified. The company must look at the composition of the products and chemical constituents. Here companies typically are surprised, and it is often necessary to substitute materials and chemicals. Materials will need to be substituted if they have either environmental or health-damaging effects or are not suited for recycling for other reasons. Materials that do not hold a recycling potential or contain harmful chemicals are abandoned in the Circular Economy. Even though REACH (EU source) offers restrictive chemicals legislation, companies still have a lot of responsibility in knowing the compounds of their products. The new EU regulation on extended producer responsibility (EPR) on products implies new responsibilities of the company. It is a good idea to be at the forefront of legislation, because it gives the competitive advantages to show the way for customers and other stakeholders.

In many industries there is an increased focus on chemistry. This is true also in the construction industry which historically has some ugly cases where chemical content shown to have harmful healthcare effects (as: PCB, asbestos, and others). As houses become insulated and airtight to harvest energy savings, indoor climate challenges from the gasification of new building materials will appear. Here there is little regulation today. Architects and Asset Owners have begun to focus on this due to an increasing use of sustainability certification schemes on buildings (LEAD, BREEM, DGNB, etc.), and the following eco-labels (Environmental Product Declaration) for building materials, and Product Environmental Footprints (PEF) in general. It is important that manufacturers focus on chemical content because this topic will raise in awareness. Companies must avoid hazardous chemicals that pollute the surroundings, and the material flows, and this is also a focus in the legislation on Circular Economy. Examples of new value chains for the most significant materials such as packaging, plastics, textiles and building materials are shown in Chap. 9 together with examples of new business models.

For many industries, it is also necessary to look at the packaging materials. In particular, the food industry is affected by new legislation on single-use packaging, as well as the EPR that will be introduced on packing material as of 2025. This means changes for the recycling of plastic and other types of packing materials in the coming years. Just as the reuse of packaging, especially secondary packaging, is reintroduced because there is a large waste of packaging. In some industries, packaging will make up a small proportion of the product impact during the life cycle, but anyway there may be good gains to be made.

C. Mapping of cash flow in the full value chain

When the overview of the full value chain of the products and materials is in place, cash flow must be linked to the product value chains. To design a new business model and to calculate a business case, it is necessary to know the cash flows of the materials and the products in the full traditional value chain and in the circular value chain. Interestingly, this mapping of products, materials, and cash flow will often crystallize innovation and business opportunities for management in new ways. It will become clear what recycling and reuse brings and what challenges the linear business model holds. It becomes visible where the new potential for competitiveness and profitable business lies in the new value chains that occur.

Of course, management already know the elements of their business. Nevertheless, new business opportunities almost always arise when management dare to think out of the box and embrace the full value chain in a new business perspective in a time where responsibility expands, material loops occur and the cost for disposal switches to an income for resources. In visualizing the value chains companies find business in rethinking their products in the Circular Economy. This is exactly where the great innovation potential for new business models arises and where the exercise of developing a circular business model becomes fun. This is why future strategy consultants and board members need to fully understand the Circular Economy. The changes that particularly drives innovation from a linear to a circular economy are:

- Pricing of resources rather than waste. Waste drives cost to dispose, whereas the future will make the recycled resources hold a new income stream. Understanding these new cash flows are important. Today it typically costs money to dispose waste. In the future, business and consumers will receive money for their used resources if they are well sorted, tagged, and ready for recycling.
- Costs and investments in new recycling technologies and new common material loops is put on the products with the EPR. An environmental fee or additional cost will be put on the materials and product to facilitate investments in recycling. The fee or cost will be determined based on the level of recycling, where recycling and recyclability will hold the lowest costs/fees.
- New market positions with new consumption models to ensure reuse and recycling, facilitates new price models for the products and services provided in the future Circular Economy.

A lot of energy is spent in literature and in acceleration programs in re-designing products for the Circular Economy. But it is often when material flows and cash flows are mapped that the actual innovation occurs. Of course, there is a need for redesign of the linear products. But first and foremost, the business models and the material flows must be understood in the new value chains. The new cash flows that come out of resource prices and recycling of resources is the innovation points, especially when coupled with digital technologies to ensure traceability and transparency.

D. **Design and describe the new business model and redesign the products (TO-BE)**

Now the task is to create the new full value chain and business model in details for the Circular Economy. This often requires more iterations and collection of data from the company and external sources. Again, the principle of Circular Economy sets the bar for development of the new business model, and prolong lifespan, reuse, maintenance, and repair often holds overseen potentials. Think New! It must be clear where the value creation is and which part of the value chain that the company wants to own in the future.

Future Products and Solutions
Working actively with steps A to C in this transformation model, the new business models emerge. Based on mapping the product value chain, the new material flows, and the cash flows. It is important to make every effort to visualize the new business model, throughout the full value chain. Then it becomes clear where the company's new business potential arises and what part of the value chain the company should own in the future. This visualization is of great importance in understanding and communicating the business models of the future. It is important to spend a good amount of energy playing with the visualization and to be careful in the presentation of it.

In the description of the new business model, it is important to look at products and customers in the context of a sustainable and circular future. Many customers, whether B2B or B2C, do not necessarily know the answer to the future products or the future solutions. The traditional methods such as user-driven innovation fall short here. Companies themselves must dare to propose the new solutions based on the disruption potential of the sustainable transformation that becomes clear by the value chain mapping in the earlier steps of this process. It may still be relevant to involve customers and salespeople, but the company must dare to invest in new solutions and new products if they want a first-mover position.

All challenges with existing materials and products were listed under step B and are to form the new technical requirements for redesigning the new products. Often, companies can redesign the products themselves, and sometimes specialists typically carrying technical knowledge and material knowledge are needed. Therefore, it is important that material experts and CE-experts are involved already in the mapping phase since they then can guide the company in the right direction. Thus, the most important strategic partners and advisors for the company will be the nerds—the technical nerds and the material nerds, and the traditional business and management consultants will become secondary. Unless you can find advisors, who possess both types of skills, and more of these will occur in years to come.

Developing a Circular Business Model

It is important to include customers and suppliers into the new business model but to view them in a new light. In the sustainable future, customers will hold the materials that manufacturers need. In the same way, it is meaningful to look at new types of suppliers and new customers in the new circular value chain. There's nothing new about this process—it's a completely traditional strategic thinking, but this inside-out mindset was lost a bit in the quest of making the customer king. Again, customers, customer segmentation, and marketing should still be thoroughly elucidated in the business plan, but the company needs to understand the products, materials, and material loops before creating a credible narrative about genuine, well-documented, sustainable products.

If the company chooses a first-mover position, it must dare to take wrong turns and dare to try out new solutions. Solutions that do not necessarily work on the first try will contribute to the learning of circular products in a Circular Economy. The new circular business models have already crept into the digital world. Device software is updated as improvements come, without disposing of the device. This is the mindset that we need to bring to our physical products. We need companies to build on the existing products instead rather than make customers throw things out whenever they are intrigued by new functionalities, or think they experience a new need (3).

Once the new business model has been drafted, work begins on redesigning the products for circular loops. Adherent to the circular principles—see above. It must also be documented that the new products have a more sustainable footprint (via LCA) than traditional products. Here the focus must be on both Circular Economy and on climate impact.

Company and Customer Business Case

E. When the new business model is ready, it is important to calculate the Business Case, including the customers' Business Case. Only then it becomes clear whether the new business model offers both an attractive environmental footprint and an attractive business potential for the company and the customers. This is often the case and not really a surprise, as value is retained in the products rather than disposing of them. Often it is possible to make a profit from this retained value.

It is important to include different price scenarios and sensitivity measures in the Business Case due to the expected changes in material prices, disposal prices, and product prices will change over time in the new Circular Economy. It is important to set various assumptions for these scenarios and get an idea of the price elasticity of the products—also as recycling percentages are being introduced and material types are being phased out by EU legislation. This, in turn, requires knowledge from industry organization and Circular Economy specialists who dare to make suggestions on how these changes will occur. Expected taxes on carbon and recyclability will enhance this phase of the Circular Economy but at this moment, the overall framework is lacking.

Preparing the business cases is not an easy exercise, but certainly worth working on in depth in the preparation for a Circular Economy. Many companies are also experiencing rising raw material prices and uncertain supplies, which also must be included in the business case, and in this way business cases will not look like they used to (3)—the future offers new types of business cases.

Business cases must in the future include the company's non-financial gains and environmental impacts, and more companies incorporate this into both internal and external reporting to support the sustainability goals the company is working towards. The new CSRD and ESG-taxonomy will speed up the focus on other than financial data.

F. Identify existing and new take-back systems.

With a new business model and the new requirements to product redesign, the company must identify suitable take-back systems for the new products and materials. It covers both investigation of the existing systems as well as future new return systems and material loops. In most European countries it is not allowed to reinvent take-back systems if existing well-functioning systems already are in place for a specific material or product at national level. Volume of uniform, recycled material is crucial to create profitable take-back systems. The recycling infrastructure and new technologies are drive by access to volume because the systems for handling of "waste" or resources are most efficient on a large scale. Thus, it is important to look for existing take-back systems on a scale that can or should handle the products and materials in the future. It should also be investigated whether some of the recycling companies specialize in the specific product group or specific material loop and find the right strategic business partners here.

The recycling industry often possesses the expert knowledge for handling recycled resources. If a company fall short of existing return systems, it may be interesting to work with competitors through trade associations or other trade organizations or otherwise to promote a circular flow of material on an industrial level. The breweries in Denmark did this when they developed ***Dansk Retursystem*** (Danish Return System) with a deposit on bottles. They have achieved recycling rates of more than 90% on glass bottles, plastic bottles, and cans. EU has decided on a uniform deposit system for drinking bottles, but implementation is lacking.

Many trade associations and companies are in the process of designing circular systems for selected materials, such as plastic, organic fractions, etc. This is mainly due to the Extended Producer Responsibility that will be introduced on the years to come, or situation where the existing EPR will be sharpened including the responsibility and the financing of the full value chain of all types of product categories.

If there are no return systems available, do not be deterred not even as an SME. There are interesting business opportunities in entering new ground and contributing to influence the systems of the future. It is always important to look at systems in other countries. They may have good systems that are worth copying. This is also the basis for entering new corporations in the value chain. It is important to understand that in small countries, there is often not sufficient volume available for many of the material loops needed at national level. In these cases, the material loops and return systems need to be built on an international or EU level. This again can be done by relevant subject specialists, organizations, and CE-experts.

When identifying a (new) take-back system, it is important to gain insight into the potentials of the digital technologies. The future will bring labelling of products and materials so that they can tracked and traced throughout the lifecycle to create clean material loops. Already now some products are subject to mandatory Digital Product Passes, as for example building materials. The future also offers the framework of the PEF (Product Environmental Footprint) that will unify the way impacts and recyclability is measured and traced.

G. Life Cycle Analysis, Documentation, and Certification

When the products are almost completely re-designed and the business model including take-back system has been mapped, it is important to create credible documentation of the product's environmental impact. The documentation should be based on the principles of lifecycle analysis (LCA) because this will become a requirement for documenting sustainability. Over years, the EU will introduce Sustainable Product Initiative on all product and standardize the LCA methods and develop the PEF (standard Product Environmental Footprint), described earlier in this book. This is to counteract greenwashing and false communication on product sustainability. The implementation of PEF will still take some years.

The lifecycle analysis (LCA) may be benchmarked against traditional products to demonstrate lower impacts. Many sustainability parameters, circular economy, and climate footprint must be documented. The new EU taxonomy will also become the standard for product just with more impact measures and sometimes up to 16 sustainability parameters. Of course, not all these parameters affect all products, but more than just climate impact must be measured, as legislated in the new EU taxonomy on ESG.

New public and private sustainability ecolabels are currently emerging due to the need for transparency, traceability, credible and impartial documentation on a market filled with greenwashing. It is a big challenge to find the certifications and documentation that will build credibility for the circular products. Analysis show that consumers are currently losing confidence in ecolabels. Probably because too many certificates have promised too much, and because the marketing on their achievements has been misleading. Consumers are beginning to discover this, and many talented consumers and NGOs, address these challenges of empty or erroneous marketing. These cautious stakeholders are often influencers on the SoMe, so it is important to look deep in the criteria of the labels that the company choses.

The jungle of ecolabels will only increase in the years to come, and it requires insight to choose the right labels and documentation. As mentioned earlier initiatives and standardized ecolabels are on their way from EU and all ecolabels must be mapped to PEF. So, it is important to ensure in the choice of labels that the labels in mind are coherent with the future EU regulation and the EU-PEF for the specific product categories. To amend for this, other companies in the same industry, industry associations, or material experts should be consulted for help with the choices on building the DPP. Just as it is important to be clear on what is documented with the certificates invested in. It is important to realize that most ecolabels are private or semiprivate. It may be difficult to see how they are financed, how often requirements are updated, and how they update their criteria. As well as which requirements the labels set to the assessing bodies (those who evaluate companies and products) is not always transparent. The quality of the brands at all levels is very variable. Each industry and each product group have its own standards and its own brands.

> **When choosing an ecolabel or certificate the following is important:**
> - Transparent assessment criteria are publicly available.
> - Traceability is part of the criteria and evaluation.
> - LCA-methodology is the basis for assessing the footprints and the circular impacts needed.
> - Footprints involve evaluation of other impact factors than GHG (CO_{2e}). Look at the ESG taxonomy.
> - Chemistry becomes more and more relevant as a parameter and should be included.
> - Security that the ecolabel will be compatible with the PEF criteria to avoid doing the work twice.

In choosing ecolabels, it is important to distinguish between product labels and company labels. In many respects, product labels will be sufficient. In some cases, there may be requirements for sustainability certification of the entire company, which often includes framework tools and annual, systematic measurement of the company's progress. The EU are standardizing the methods for assessing the company impacts in the new EU-OEF (Organizational Environmental Footprint), where assessment of the full company is standardized, which will affect certification of companies in the future.

Another applicable method that is increasingly used to document the sustainability of the company and its products is the strategic tool Sustainability Roadmap where the framework of ESG or the Sustainable Development Goals (SDG) may be selected. Here the company includes targets, actions, and policies based on the Materiality Assessment that has helped identify the significant footprints. This Sustainability Roadmap is a tool to create progress and meet the monitoring requirements of the CSRD legislation. Hereby the company get an overview of the strategic roadmap which is also good for communication and prioritizing. It is a systematic way of getting started, instead of just throwing themselves into a few details and actions. These kinds of strategic sustainability tools are increasingly used and are a pragmatic and easy approach towards sustainability and still help to prioritize and act where it makes the most sense for the company and the planet.

Company Labels
There are many certification and reporting standards for corporate companies—such as, Global Compact, GRI, Science Based Targets, EMAS, ISO, Cradle2Cradle

and many more. Now the new EU ESG-taxonomy will set a new standard for company reporting. The certifications have different purposes, different audiences and different focus and contents. Just like the product ecolabels. It is important to uncover one's own needs and stakeholder (customer) requirements. Perhaps it is necessary to conduct a broader stakeholder analysis of what stakeholder's demand—especially in larger corporations. This includes customers, future employees, investors, etc., before choosing a scheme. But for SMEs and many others the mentioned schemes are very comprehensive and applicable in a convincing way that also can be used for internal and external communication.

Business Plan, Implementation Plan and Communication Plan

H. Implementation of any new strategy requires a business plan, an implementation plan, and a communication plan. This must include a market analysis preferably also with customer involvement in a traditional outside-in perspective based on the new products and the new business model. It is always important to have an outside in view on the company and its products. Companies that understand the challenges and seeks sustainable solutions will find that many are interested in attending a dialog on the Green and Circular transition because everybody wants to learn and to contribute to the future solutions and a better planet. Companies may experience that they get access to potential customers now in a way that they were not able to before.

The Business Plan must include customer potential, competitor map, pricing, sustainability impact and marketing efforts, as well as the need for investments and capital. There is often a need for investments in new market channels, new market communications, product development from prototype, production equipment, cost for ecolabels and certificates and other. There is an ocean full of useful templates for Business Plans. SME can get help for the transition by various accelerator program or likewise, possibly also based on public, soft money. Such programs will typically make the relevant formats and tools available for business plans and implementation plan etc. Many companies will also need this business plan to attract new investments to finance the transition.

It is also important to realize that implementing a green and circular business model is a long process. Experience shows an identification phase (A to G) of approx. 6 months depending on whether prototypes are to be developed and tested. Then there is an implementation phase of at least one year before the business model and the new products have been fully introduced to the market. The time of market penetration can vary - depending on how professional the company is, and how many muscles and money the company has for marketing. This requires a thorough and detailed implementation and communication plan. The Sustainability Roadmap is a good basis for communication internally and externally.

It is still difficult and a long process to introduce a genuine sustainable product on the market, as many customers—especially large corporations (private and public) are bound by existing contracts, traditions, and history with their existing vendors. It is time consuming for these purchasing decisions to find their way through the organization silos, and to ensure that sustainability requirements are set already by procurement. In particular, the public sector is foot-dragging in this area in many countries. It still is as if a lot of political talk never reaches the public procurement departments or the legal department. Not all the political promises for the sustainable, public procurement power to drive the green transition has yet become reality.

Most of the public procurement is still based on the initial price of the product, and thus cheap linear products beat long-term sustainable solutions. When introducing the EU-PEF, the European Commission announced that they will dictate that PEF becomes a requirement in public procurement, and that public companies and institutions only can purchase from the best half of the products in a PEF benchmark.

Communication and Education

An important element in the green transition is a good communication plan and an education plan – internally and externally. It is perfectly okay to communicate to the outside world when a company starts the sustainability journey. It is certainly not necessary to know all the answers or the precise targets, on the contrary. Companies must clearly communicate their goals and visions in good time before having achieved them and be open about the challenges in their value chain. In that way a company can invite its stakeholders on the journey towards sustainability, but they should NOT over-market or greenwash on this journey or when goals are achieved. It backfires so badly if companies are caught in undocumented green claims, and it challenges the trustworthiness of the company in the market and the harm will be long lasting. See more on greenwashing and recommendations in Chap. 5.

When the companies clearly communicate their changes, challenges, and successes they open the door to true storytelling. Marketing is hungry for real storytelling even if it seems as if many of traditional PR agents have not quite grasped the difference between real and fake storytelling. It is difficult to find marketing and PR agencies that can create a trustworthy and in dept story about the sustainability journey and many of the old mastodons are unfortunately still preoccupied with the reverse task, to build a story around an existing product in a green way = greenwashing.

No one is expected to solve all the challenges at once and it creates credibility to communicate openly on the journey. Green storytelling without doing anything new or just cutting 10% off company energy consumption is not sustainability. The new digitally talented customers and influencers will soon look through this and fortunately, a lot of innovation is sprouting with new young, digital agencies that understand the green agenda. Perhaps there is also a generational challenge in the PR agencies as well as in the companies.

Circular Economy in the Crystal Ball

Circular Economy and transition to new green business models is still full of uncertainties. Many companies choose to wait and follow how the various material flows and return systems evolve and gain more certainty in future solutions. Then the company can transform based on established solutions and technologies to well-known markets.

There is a competitive advantage in being *"First mover"* in a new and open marketplace. In a "first mover" position the company can influence the new market, on how the market develops and gain access to new strategic customers that would not be accessible in an entrenched market. Experiences have shown that even small companies can open new market segments with their circular business model and their sustainable products that they due to their small size would not be able to target because of too strong competition from large companies. Now a company suddenly offers the solution of the future, and everybody is looking for this, also large strategic customers.

There are countries in EU that want to create a green front-runner position for their businesses, such as the Netherlands, France, and Denmark, with different initiatives and a good market dynamic for green solutions, here good opportunities to test new solutions exists, also with public funding. Both for development, demonstration, and export. This to create new export opportunities for companies by having the opportunity to market and to mature products on the local markets through support and network. There are also some international collaborations to help companies market their sustainable products, both in US and Asia. Examples are C40 (cities sharing solutions), Circular Economy Fairs, Clean Tech Fairs are just a few and new ones pop up all the time. Trade associations often have focus on assisting companies to create an overview of future solutions and to promote future solutions and to promote "first movers" that others can reflect in and that are door openers in export promotions.

The process outlined in this book on transforming a company to a circular business model is largely aimed at manufacturing companies, although the mindset on Circular Economy can be used for other types of companies and entrepreneurs, as well. There is a potential in offering new, digital business models based on the products of others, where the ownership of the product is kept by either the manufacturer or the consumer. Here a circular mindset on documenting environmental impacts is equally relevant, as for the manufacturer.

In the future, large manufacturing companies will be disrupted by the green transition, and large players will fail because they are not prepared for the Circular Economy. The German automotive industry is probably the best example of companies that opened the market for disruption by holding on to old technologies and forgot to prepare for the future. They have even resorted to fraud, to greenwash and market false low environmental impacts. The transition of the automotive industry has been kicked off by one single entrepreneur—namely Tesla. These impressive and well-functioning electric cars that set a completely new agenda in the car industry, where first Nissan and since all other brands had to follow suit. This open marketplace of EVs has already been taken by Chinese automotive producers and new players like Tesla also on the European and American market. This is an excellent example of disruption of an industry that was asleep, and which was governed by financial interests of the oil industry to a larger extend than by customers or market trends.

References

Haar, G. (2024a). Rethink Economic and Business Models. *Rethink Economics*. : SpringerNature.
Haar, G. (2024b). *Nordic Case Collection*. SpringerNature.

Chapter 17
Developing a Sustainability Roadmap

This chapter introduces a tool for companies to build a Sustainability Roadmap based on the legislated process for ESG reporting. When starting the ESG reporting process a company needs to verify it's full value chain as input for a Materiality Assessment (MA). The MA is the basis for the company's decisions on the change that they need to drive and a paragraph on Materiality Assessment is earlier in this book. The EU Green Deal as well as the legislation and the EU-reporting requirements on sustainability (ESG) are based on the Sustainable Development Goals (SDGs) and it may freely be decided how the companies build their framework of managing the changes they need to drive.

According to the EU legislation (ESRS) the company must provide a plan that includes:

- Policies
- Targets
- Actions
- Allocation of resources in and outside the company
- Monitoring ESG issues raised inside and outside the organization,
- Describe how development of competence throughout the full organization is managed,
- Monitor how management is an integrated part of ESG and sustainability of the company.

The tool illustrated in Fig. 17.1 provides a systematic approach on registration and managing the above (Haar 2021). This tool will be found in a digital form as part of a new ESG platform provided by Center for Circular Economy.

Using the SDGs or the ESGs as strategic goals enables companies to create the changes for contributing to a regenerative and climate neutral value chain with positive impact on the social interactions, as well as develop their business models and their products sustainably. The strategic approach required in the transition to a

SDG or ESG parameter	Targets / descriptions	SDG or ESG parameter	Status today	Baseline	Project Owner / Project Manager	Project Resources Estimate	20XX	20XX	20XX	20XX	20XX
Insert the select SDGs or ESG parameters	Describe in details SDG targets or ESG descriptions	Describe the SDG target or ESG parameter in the context of the company	Map to the existing	If available link to baseline in numbers	Project owner from Executive Management. Project Manager in the organization	Estimate the resources needed - internally and externally. Out of pocket. Time estimate					
 7 (Affordable and Clean Energy)	7.2 Before 2030 the share of renewable energy in the globale energy mix must increase significantly. 7.3 Before 2030 the speed of global improvement on energy efficiensy must double	Has on the agenda for some time to have an energy report prepare	Has not yet had the potential uncovered	Potential not known but is estimated to be significant	Project owner: CFO NN. PM: engineer XX from production	Build a GHG baseline. Energy Optimization	Install solar panels (PV)	Climate neutral in scope 1+2			Climate neutral in scope 3
 12 (Responsible Consumption and Production)	12.4 A environmental responsible handling of chemicals and waste in the full life cycle according to international agreements. 12.6 Companies, especially corporate transnational are encouraged to use sustainable practice and to integrate informatin on sustainability in their reporting								Implement a circular business mode		

● Targets ● Policies ● Actions

Fig. 17.1 Sustainability Roadmap

Green and Circular Economy holds great business potential. Another book by the same author provides a tool to prioritize the SDGs in a strategic way and there is inspiration from companies using the SDGs as strategic drives in a Nordic Case Collection (Haar 2024a) & (Haar 2024b/24).

> **Preparing for the Sustainability Roadmap to meet EU-ESRS requirements:**
> 1. *Read and understand the SDGs, targets, and indicators (or ESG)*. The SDG framework provide very good educational material online, including small videos on all 17 SGDs in almost any language. Participating in education on sustainability and the green transition may be a part of becoming ready for transforming the business.
> 2. *Identify the company's full value chain* from primary production (raw material) all the way through to disposal and recycling of the products. The framework is the UN Climate Protocol and the 3 scopes, and the ESRS legislation (EFRAG) will provide standard value chains based on industry codes.
> 3. *Study the 5–6 most important SDGs (ESG) for the company*: Companies in the Western world need to relate to SDG 12, and many also needs to investigate SDG 7 for reduction of energy consumption and converting to renewable energy in their own buildings (scope 1 + 2). See book 2 for more information on the relevant SDGs.
> 4. *Investigate whether other SDGs (ESG) are particularly close to the core business* and the company's products. Once a company has considered the few important SDGs (ESG topics), it should orient itself in the other topics or goals. Especially among the more visionary and tactical world SDGs, as there may be one or two topics or goals that are close to the core business and products. Choose only those SDGs (ESG topics) where the company can make a significant impact through its core business, value chain, and through its products or services.
> As an example, if the company supplies water technologies, products within the fishing industry or is in the plastics industry, SDG 14 (marine life) must be investigated. If a company relies on or supplies agricultural products, SDG 15 (Life on Land) should be addressed, and so on.
> 5. Conduct the *Materiality Analysis*. See earlier here on the Materiality Assessment.
> 6. *Select a total of 3-5 SDGs (ESG topics) as the basis for the Sustainability Roadmap*. If the company is to embark on the transition to the Circular Economy and develop new business models and products SDG 12 is unavoidable.
> 7. *Describe the status of the company (as-is) today* of the selected SDGs (ESG). Identify if the company can build a baseline for the selected topics, as for example the GHG baseline or work-related accidents as of last year. If not, then the first target to be set is how and when the company has decided to start monitoring.

8. Then the targets and actions for each of the selected SDGs (ESG) must be identified and this is best done a series of workshops with management and relevant stakeholders in and around the company. This is an extensive work, and this is the most important work to be done in when a company embark the journey towards sustainability. The template below is the tool that can be used to manage and document this work. The format can be adapted to the company's own methods and presentations to match how the company usually communicates.

There are different types of targets and actions that must be included in the roadmap:
   a) *Monitoring the non-financial data* needed is also an important part of the actions needed, as many companies today are not able to account for the impact.
   b) *Targets of* climate neutrality, circularity, biodiversity, social impacts, anti-corruption etc. must be set for the selected topics or goals.
   c) *Each target must be followed be actions to drive the change* and it must be described how management is involved and informed on ESG and on the change that the company is driving. The company should include ESG reporting on baselines, actions, and achievements in the monthly management report that most companies prepare, and this is the most convincing way to document management involvement and then the referendums from management meetings is part of the monitoring process.
   d) When each target and action have been decided upon it is important to *describe the policies of the company* on these selected topics. These policies are important for future communications internally and externally.
   e) On each action it much be decided and described who in executive management will become the *owner of the action* and who will become the *project manager*. Resource allocation to drive the actions also must be described, internally and externally if making use of advisory services or other experts to drive the change.
   f) A very important target and action that must be included *are the development of competences* necessary to drive the change and to prepare the company for a Green and Circular Economy.
   g) *A monitoring tool* is needed to register and address all the ESG topics raised in and towards the organization.

The framework for the tool may be the Sustainable Development Goals (SDGs) or the ESG topics from the EU sustainability taxonomy. Many companies choose to use the SDGs as their strategic framework and then the roadmap should be built against these.

In the mapping of the value chain, the company must look at all impacts it has on people and the environment throughout the value chain and with all three scopes, as described above. Then the company wors with the topics relevant within the full span of control and set new requirements to suppliers and customers to be able to meet the targets set in the companies Sustainability Roadmap.

The topics or SDGs that the company chooses must create a long-term, measurable impact and must have strategic importance. It is not enough any more to do a little philanthropy on the side or a little for the CSR report. ESG or SDG are new market conditions and new business opportunities, and it will be the most important strategic management topic for many years. ESG or SDG have the same strategic weight or more as other business decisions. The targets and actions set does not necessarily have to be measurable impacts on a national or global level, but it must be significant for the business of the company and its size. The company must dare to risk and invest, because ESG/SDGs matter to the business in the long term.

SDG or ESG must:
- be initiated by management (executive management and/or non-executive board of directors),
- changes the way the company is doing business,
- result in investments to drive change, and traditional business case analyses and other relevant market assessments must be conducted fully in line with other strategic management decisions.

Greenwashing has been a problem for many years, also with the SDGs. This is done by mapping existing activities to as many SDG as possible, to be used in communication to customers and stakeholders without conducting any changes. Then the SDGs become a play of communications cards and does not drive the changes that the planet needs. This is a lack of respect to overall objective with the SDGs (ESGSs) and the wording of the targets here. It is perfectly acceptable not to solve all the problems at once. Companies, on the other hand, must assess and build knowledge on their impact on the planet, nature, and people. They must be open about their negative impacts as well as the achievements they create and include this in their roadmap and their communication.

SMEs Versus Global Corporates

Experiences in Denmark have shown that SMEs are very innovative—even more than many global corporations trying to articulate the SDGs and sustainability. SMEs are feeling great pressure from the market these years to switch to green and circular business models. The price pressure from low-price products has become heavy and new types of business models are in need. Many SMEs within manufacturing are looking for new business and are ready to innovate in depth. The global corporations should focus more on collaborating with SMEs and elevating their solutions. International corporations must dare to take the risks to choose new, innovative suppliers, even if they are smaller and more uncertain suppliers than they are used to. The innovative power of SMEs will only be implemented if customers dare to choose new solutions and dare to support the scaling of them. Then these new solutions may become an important part of the global corporation's own transition.

The large corporations have a strong execution power in the existing markets, but they are often limited in innovation power and agility in implementing new solutions. The global corporations are highly organized in silos, and the too short-term management goals overshadow the opportunities rethink and to create change across the company and for the planet. The green transition requires long-term strategies and investments. The long-term investment and business models are challenged by the fact that corporate CEOs often only hold their positions for a few years (4-6 years) and their performance measures are often creating short-term shareholder value. In addition, the natural human fear of changing something that has worked for many years is an obstacle for the green transition.

It will be production companies, product owners, and those who own the consumer data who will own the future marketplaces. Everything in between, as distribution and retail will be subject to severe disruption. This is also the ground on which many of the major IT corporations like Amazon, Google, and Facebook are building their business models. In that understanding, much digital disruption will occur, and this is already booming. In the future, disruption of production industry because of the shift to the Green Economy will arise, as well as big corporations fall because they were not prepared for the Green and Circular transition.

All Are on a Sustainability Journey

Sustainability is a journey and a long and never-ending journey. A journey for everyone—governments, citizens, and businesses. It's okay to take the actions on board that make sense for the business at an affordable pace. Leaders' lack of awareness of the impacts of their business in the full value chain is a much larger problem than not being able to solve all the problems at once.

The Sustainability Roadmap must of course be adapted to other business tools or methods that the company normally uses for implementing strategy and larger projects. The Sustainability Roadmap is also a good communication tool, both internally and externally. Not that it needs to be publicly displayed in detail, but in an overall summery version. Most important is to maintain an overview of the company's significant impacts and to keep track of which efforts have been selected and which have been deselected based on insight and accountability.

Many alternatives exist to approach the SDGs or ESGs and many are available online. It should be investigated before implementing a tool if it provides for both scope 1+2 and scope 3.

A company cannot be expected to save the whole world at first step or embed all actions at once. It is also well known that not all solutions are available now. A company is expected to have an awareness of impact and level of footprints. The Materiality Assessment creates this overview and awareness of the company situation. The Sustainability Roadmap is required by the CSRD and is the basis for a good communication platform, internal and external. This can also be mapped to the Sustainable Development Goals. This Roadmap is the start of a long journey for every company, and nobody expects that all is solved at once and that all the solutions are in place. Management is expected (also according to legislation) to know and understand impact of the company and the extended value chain on all ESG parameters, they are expected to know the reasoning behind the prioritized actions above other actions in a sensible way, and they are expected to publish a plan for how to minimize impacts and build regenerative ecosystems and business models that support this. All this is also stated in the legislative requirements of EU. See Chap. 6.

It is not the destination. It is the journey.
— Ralph Waldo Emerson, author, and philosopher

References

Haar, G. (2021). *Sustainable Development Goals as the most important plan for companies (danish)*. Quare.
Haar, G. (2024a). Rethink Economics and Business Models. *Rethink Economics.* : SpringerNature.
Haar, G. (2024b). *Nordic Case Collection.* SpringerNature.

Chapter 18
Summary

Our planet and the economies are in a shape that calls for dramatic change. The Sustainable Development Goals (SDG) have become an important map to visualize and teach the actions needed on a global scale.

Human spread on the planet and the unsustainable human activities has put the planet into The Anthropocene Age where climate impact, land use and resource scarcity are affecting each other negatively. We need to redirect this wheel and create a sustainable living, regenerative ecosystems and protect the climate to create a sustainable and fair living for all humans.

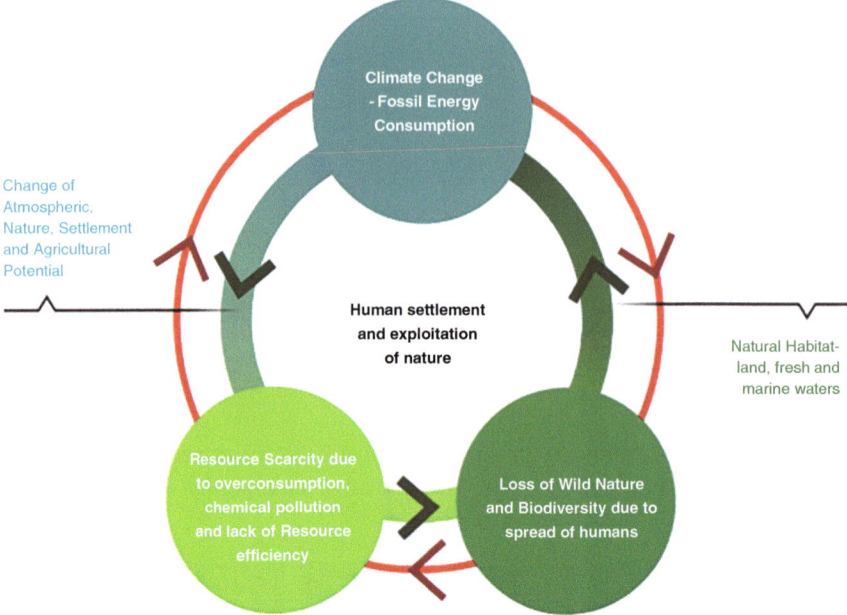

Extreme weather with increased and more precipitation in some areas and drought in other areas means that climate adaptation is required in all countries on earth, at all levels of society. Temperatures are rising and the melting of the ice has started, thus the water levels will continue to rise. All this already results in pressure on food systems, cities and human lives and will cause climate refugees around the planet, especially from the regions around the equator is a reality because these areas are becoming uninhabitable.

Climate neutrality and a sustainable way of life is at urge if we are to meet the Paris Agreement, and to avoid catastrophes. The great transition to at Green and Circular Economy is a necessity and illustrated here as the Climate Nexus:

- The global energy supply needs to become renewable,
- The global primary production and use of land and wild nature needs to be sustainable and based on regenerative ecosystems,
- Western World consumption must be based on a clean and Circular Economy, and
- Global transport to sustainable and collective forms, mainly on electric power.

18 Summary

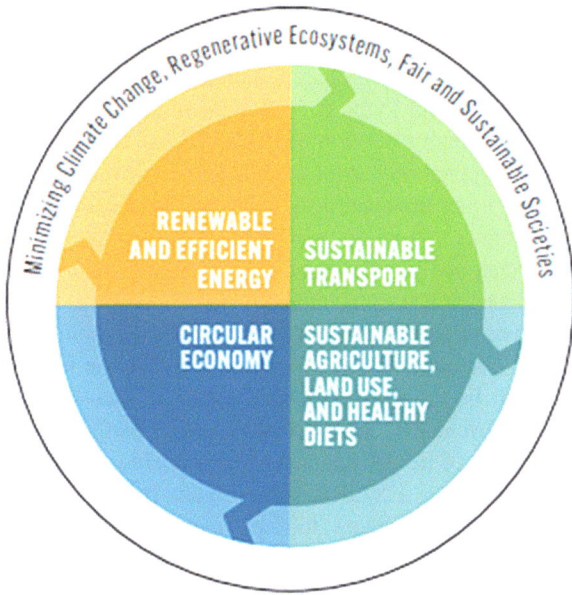

These are massive changes for all, and a holistic and intelligent green transition can lead to a genuine sustainable human living necessary to provide for 10-11 billion people on Earth by 2050. The Climate Nexus and climate action can become the train for a creating a sustainable and fair way of living for the benefit of nature and people.

Companies are the wheels of transforming our society and our economies. Management and business owners are now subject to an extended responsibility in the extended value chain (scope 1 + 2 +3). Companies on the EU market are meet by a lot of new legislation under the EU Green Deal with demands on transforming their value chains, business processes, products and reporting to meet the new inner green market in 2030. For this they need tools, competences and sharing experiences. This will be the largest transformation consumers and businesses will meet in their time. This book builds an understanding of the burning platform for change, describes the future that we need to form with the great transition to a circular and green economy and provides tools to manage the transition in the companies.

With a broader understanding of the link between climate impacts and sustainability, everyone can contribute to making holistic choices. Not only by switching to renewable energy, but also by transforming to efficient and sustainable transport, to Circular Economy, and convert to regenerative ecosystems and food production on land, in oceans and in fresh water. We can create a truly sustainable world if we want to. The green transition is complex, but with an intelligent approach we can reduce the man-made climate impacts and achieve a lot of derived positive effects. Understanding this Climate Nexus is important for companies when they make choices and actions for their extensive work on sustainability. Business leaders need to understand the consequences of their choices in time and in the full value chain.

Unfortunately, not many easy solutions are left, apart from the rapid transition to renewable energy. The Doughnut Model, presented in the (Haar 2024), is a model for creating sustainable development within in the planetary boundaries, and still create growth and decent living for many. In this way, the climate transition becomes a train for sustainable development. Capital and the financial sector have a responsibility to set sustainability requirements with their investments.

Pension funds and other financial institutions that manage citizens' money need to look at infrastructure solutions for renewable energy, sustainable transport, and the transition to a clean and circular economy of manufacturing and food production. Otherwise, citizens will start by choosing other solutions.

Circular Economy is a significant part of the great transition to a new economy. Companies are very central in the transformation to a Circular Economy, and genuine sustainability will be the most important strategic agenda for companies for many years to come. Circular Economy is addressed in SDG#12. The biggest challenge in the transition to a Circular Economy for companies is that they must understand and account for their ENTIRE value chain, from extraction of primary resources, manufacturing at all stages, to reuse, recycling and disposal in responsible ways.

Circular Economy is to reuse products and recycle resources so that companies and regions become independent of virgin raw materials. The planet cannot provide the growing population with resources if we continue to throw away at the scale as we are doing it now—especially in the Western world. Therefore, our consumption must be decoupled from the extraction of virgin resources. To get started with the transition to Circular Economy, companies must be in control of the global value chains and must be able to document their environmental impacts throughout the full value chains. This requires far more transparency and traceability than both businesses and consumers have today.

Companies must design products according to the principles of the Circular Economy, as:
- Longer shelf life and providence of maintenance.
- Reuse and resale.
- Separation at material level so that materials can be recycled in cascades.
- Without harmful chemicals polluting products and material flows
- Business models that promote reuse, recycling, and clean material flows.

18 Summary

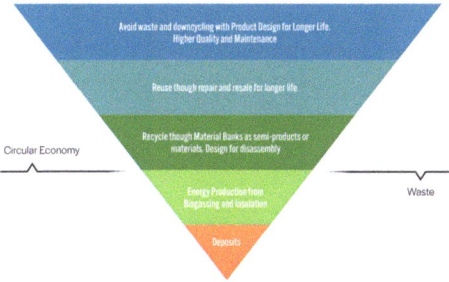

In a Circular Economy, the goal is to preserve as much value in the products for as long as possible. This is well illustrated in the hierarchy of resources.

Companies and extensive legislation (EU Green Deal) must ensure clean and segregated material flows through new infrastructures supported by take-back systems, sorting, separation, quality assurance and material banks.

Circular Economy is an important solution to counter man-made climate change, because the consumption of products in the old industrial countries account for almost half of the global climate impact and recycling of material resources itself create enormous energy savings (35–95%) compared to production from virgin resources.

The new Green and Circular Economy is driven by:

- Consumer demand and demands for more sustainable products and consequent demands for transparency and traceability to enable consumers to make a truly sustainable choice. In addition, there is an increasing demand for reused and recycled products. This is supported by the development and implementation of Sustainable Product Initiative with Extended Producer Responsibility (EPR), new ecodesign criteria, and the EU's new Product Environmental Footprint (PEF) for product labelling.
- An growing awareness on climate change and sustainability from all company stakeholders and widely among the citizens, employees, as well as an increased focus on the link between our material overconsumption and climate change.
- The EU Green Deal, where a transition to a clean and Circular Economy is one of the pillars in creating a sustainable region by; climate neutrality in 2050, where economic growth is decoupled from resource consumption, introduction of regenerative ecosystems on land, fresh water, and oceans, and where no person and no area is left behind. All described in the extensive legislation and defined in the EU Taxonomy for Sustainability based on ESG—Environment, Social and Governance Impacts.

All actors must be involved in the transition to a Greem Economy for it to succeed. Companies are central to the green and circular transition, and it requires new business models to meet market demands in the future, and there is great business potential in exploiting the value embedded in products and materials rather than getting the consumer to throw them away and creating waste. Consumers and manufacturers need to come closer together to promote mutual understanding of manufacturing processes and of consumers' wishes for genuine sustainability and proximity. There is a need for a radical change, which the Circular Economy is. The local collection and sorting of household waste and later resources is an important element in the proximity that initiates this transition.

Upcycling is not always circular; upcycling makes sense if we have lots of waste to upcycle from. The Circular Economy aims to move away from waste generation, and all products and materials must be included in circular material loops already from design and production. Therefore, both the future recyclability of materials based on recycled materials are important—i.e. both history and future.

There are no standard business models that solves the Circular Economy. Business models are individual and are fitted to the company, the markets, and the products. In the future, companies' business models must ensure that they continuously hold access to resources and that their impact on the planet is minimized, and perhaps customers no longer want to buy products, but simply borrow or rent them. Rental model, deposit systems and take-back create customer loyalty and responsibility for the full value chain of the product.

Companies need to transform their business models and take responsibility for their global value chains they take share in. This will put products and materials at the heart of companies' strategy in the future, rather than a one-eyed focus on go-to-market strategies. Thus, the strategy processes change from what business leaders are used to.

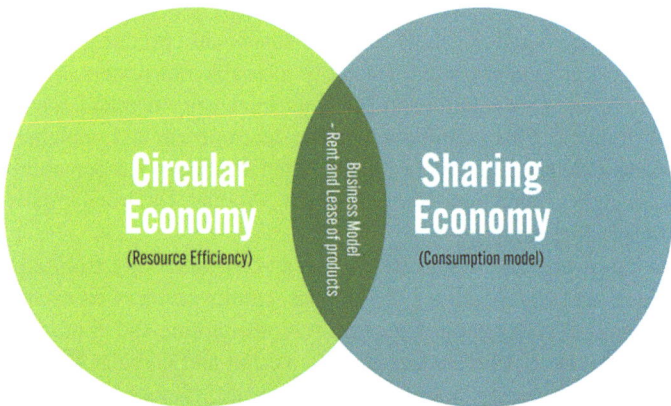

Sharing Economy and Circular Economy are not the same, although they are often confused. Sometimes a sharing economy business model is circular and reduces the use of resources and the impact on the planet, sometimes the opposite.

Circular Economy, sustainability and Industry 4.0 promote each other, and it requires new competencies also in management if the company is to survive on the marketplace in 10 years. On the other hand, there is a huge financial potential, especially if existing SMEs become an active part of the green transition.

Transforming the businesses is a strategic exercise that is based on the products and materials to deliver the sustainable products that customers demand, and that legislation requires. In the future, companies will have to document their sustainable footprints throughout their life cycle in the same way level at which they today measure their financial results. Transparency and traceability throughout the full value chain will be crucial for companies to market their products on the EU's inner green market.

Companies will experience major changes both in the market, in the supply of raw materials, from consumers and from legislation, but perhaps the biggest change facing companies is the demand for proximity in the production of things and food. With proximity and a better understanding of the value chains, traceability and transparency emerge, which are needed for companies to get the right prices for the goods and thus account for a sustainable harvest of natural resources.

For all this to happen companies needs tools, methods and roadmaps that are presented in this book and then they need to build their management skills because it is very different to manage a company that is purpose driven and not only hold financial targets but also ESG target and actively participate in the transition to a fair and sustainable planet.

Methods/Tools	Description	Illustration
Examples of Full Value Chain	Graphic representation of the full value chain and the impacts from products and materials in—upstream and downstream. The full value chain description is important for all sustainability work. Some product value chains may consist of sub-value chains—for example, food products, textile products, etc.	
Circular Business Model	Standard description of a circular value chain at material level to ensures reuse, recycling, material banks, and new business models	
UN GHG Protocol for Companies	The visualization of the 3 scopes of Company Impacts in the full value chain. This is a good inspiration for describing the full products value chain	
Materiality Assessment	A tool to identify and prioritize corporate sustainability initiatives based on the full Value Chain. ESG impacts are held against the effort to drive changes. Business risk and business potentials is added as an extra dimension in the assessment. The MA is mandatory according to EU reporting regulation as CSRD and SFRD	
Minimizing Climate Impact in Scope 1+2 – Energy Optimization and Renewable Energy	A process to minimize climate impacts in scope 1 and 2 through energy optimization and installation of renewable energy with a holistic approach	
Sustainable Building towards Impact Positive	The five elements of a sustainable building regardless of certificate or ecolabes	

(continued)

Methods/Tools	Description	Illustration
Process for transforming to a Circular Business Model	Strategy process to transform the company from a linear business model to a circular business model preparing the company for the Circular Economy. The process embeds the full value chain and includes redesign of products	
Sustainability Roadmap	A roadmap tool to document and drive targets, actions, resource allocation, and policies of the company over 3–5 years. The roadmap is a good basing for training employees as well as communicating the company strategy and actions. The roadmap may use ESG topic or SDGs as the overall navigator	

Reference

Haar, G. (2024). Rethink Economics and Business Models. *Rethink Economics*. : SpringerNature.

MIX
Papier aus verantwortungsvollen Quellen
Paper from responsible sources
FSC® C105338

If you have any concerns about our products,
you can contact us on
ProductSafety@springernature.com

In case Publisher is established outside the EU,
the EU authorized representative is:
**Springer Nature Customer Service Center GmbH
Europaplatz 3, 69115 Heidelberg, Germany**

Printed by Libri Plureos GmbH
in Hamburg, Germany